W0079296

HORIZONS OF QUANTUM CHEMISTRY

ACADÉMIE INTERNATIONALE
DES SCIENCES MOLÉCULAIRES QUANTIQUES

INTERNATIONAL ACADEMY
OF QUANTUM MOLECULAR SCIENCE

HORIZONS OF QUANTUM CHEMISTRY

PROCEEDINGS OF THE THIRD INTERNATIONAL CONGRESS
OF QUANTUM CHEMISTRY HELD AT KYOTO, JAPAN,
OCTOBER 29 – NOVEMBER 3, 1979

Edited by

KENICHI FUKUI
Kyoto University, Dept. of Hydrocarbon Chemistry, Faculty of Engineering, Kyoto, Japan

and

BERNARD PULLMAN
Institut de Biologie Physico-Chimique, Paris, France

D. REIDEL PUBLISHING COMPANY
DORDRECHT : HOLLAND / BOSTON : U.S.A.
LONDON : ENGLAND

Library of Congress Cataloging in Publication Data

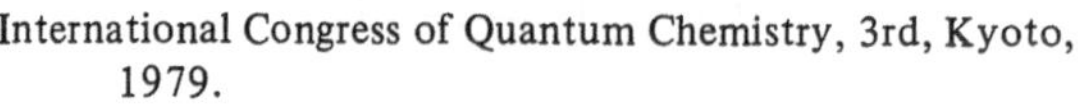

International Congress of Quantum Chemistry, 3rd, Kyoto, 1979.
Horizons of Quantum Chemistry.

At head of title: International Academy of Quantum Molecular Science.
Includes index.
1. Quantum chemistry–Congresses. I. Fukui, Kenichi, 1918- II. Pullman, Bernard, 1919- III. International Academy of Quantum Molecular Science. IV. Title.
QD462.AII57 1979 541.2′8 80-12124
ISBN-13:978-94-009-9029-6 e-ISBN-13:978-94-009-9027-2
DOI: 10.1007/978-94-009-9027-2

Published by D. Reidel Publishing Company,
P.O. Box 17, 3300 AA Dordrecht, Holland.

Sold and distributed in the U.S.A. and Canada
by Kluwer Boston Inc., Lincoln Building,
160 Old Derby Street, Hingham, MA 02043, U.S.A.

In all other countries, sold and distributed
by Kluwer Academic Publishers Group,
P.O. Box 322, 3300 AH Dordrecht, Holland.

D. Reidel Publishing Company is a member of the Kluwer Group.

TABLE OF CONTENTS

OPENING AND WELCOME ADDRESSES

SYMPOSIUM I. NEW APPROACHES IN QUANTUM CHEMICAL METHODOLOGY

Chairman: M. Kotani

SYMPOSIUM II. BOND FORMATION AND BREAKING

Chairman: M. Simonetta

SYMPOSIUM III. MOLECULAR INTERACTIONS

Chairman: J. Koutecký

OPENING ADDRESS

Kenichi Fukui
Kyoto University, Kyoto, Japan

Ladies and Gentlemen,

It was six years ago, in July, 1973, on the occasion of the First International Congress of Quantum Chemistry in Menton that Professor Daudel, then president of the International Academy of Quantum Molecular Science, proposed having our 1979 meeting in Japan. When we considered that proposal, we could not help feeling some hesitation, because Japan is located in the so-called "far east", that is, far from Europe and America where many scientists in our field live, so that even on the basis of geography alone, it seemed that Japan was not a suitable place for an international congress of this sort.

Nothing, therefore, gives me more pleasure than that various adverse conditions have been overcome and we have here now so many participants from abroad in addition to a great many Japanese participants. Such a large number of quantum chemists and scientists in related fields assembled in one place to exchange ideas about recent topics! As it was with our previous two congresses, this is a wonderful event in the development of quantum chemistry. To the participants who energized this congress and put soul into it, I wish in the first place to express my heartfelt gratitude. We are heavily indebted to Professor Pullman and Professor Parr for this: they rendered their good services for the convenience of participants from Europe and America, as well as to Professor Daudel who prepared posters distributed to institutions around the world to inform them of the Kyoto Congress. I should also like to express appreciation to the International Academy of Quantum Molecular Science, under whose auspices this congress is held.

A distinguishing feature of this congress is its co-sponsorship by the Science Council of Japan and the Chemical Society of Japan. The former is an official organization of the Japanese Government, and the latter is one of the largest scientific societies in Japan. This sponsorship elevated the Congress and brought privileges and advantages to its organizing.

K. Fukui and B. Pullman (eds.), Horizons of Quantum Chemistry, vii – x.

During the Congress, each morning is allotted to one symposium, in which three or four invited lectures are scheduled. The proportion of these invited lectures to the total contributed papers is large compared to the usual case in international meetings. The chairmen and the 17 lecturers accepting the requests of the Academy and the organizers assisted in increasing the significance of the Congress.

In financial support to this congress, we received contributions from a number of economic institutions, companies, and foundations. Without this generous help, the Congress would not have been possible. It was also fortunate that most of the symposium chairmen and invited lecturers were able to obtain support from their own sources. And Professor Löwdin kindly offered to publish contributed papers in the International Journal of Quantum Chemistry as Proceedings of the Congress. For all of these favours I am grateful beyond expressing.

The members of the Organizing Committee and the Sub-committees, the Funds Committee, and the Finance Committee spent much time organizing this congress, and we are indebted to the administration bureaus of both the Science Council of Japan and the Chemical Society of Japan for their hard work, as well as that of the Kyoto International Conference Hall. Deep appreciation is due to all of these people.

I myself am firmly confident of the significance of this international meeting, supported as it is by the good will and exertion of so many people. But whether it is truly significant or not depends also on what is contained in the meetings and what arises from them. For the cooperation of all participants through their presentations and discussion, and for the efforts of session chairpersons, I should like to express thanks in advance.

Quantum chemistry has kept pace with the development of the chemical sciences and now serves as a foundation in every area. This is a logical consequence, and means that chemistry is developing as it should. So an important question is: What position does quantum chemistry actually hold now? We can, happily, make the following statement: --- now, the chemist's dream of being able to calculate the structure and property of a given compound or the velocity of a given reaction with only paper and pencil is slowly but surely approaching realization. In fact, the pertinent question now is, to what extent will chemistry, which so far has been a science of a strongly empirical nature, attain a nonempirical character? Simultaneously, it is now time to ask, to what degree will the nonempirical character of quantum chemistry be utilized for the purpose of advancing empirical chemistry?

In these ways quantum chemistry will function both theoretically and methodologically. The remarkable recent progress of quantum chemistry is of course mainly due to methodological advancement, but it is also a result of improvements in physical measurement techniques and in high-speed computers. We are so encouraged by these circumstances as to unhesitatingly introduce quantum-chemical ways of thinking into every field of science which relates to chemical phenomena. The propriety of such an idea will be obvious if you think, for instance, what a great influence Professor Mulliken's concept of charge-transfer has exerted on various fields of pure and applied science. Thus new quantum-chemical concepts will accordingly be required to interpret future discoveries. Conversely, it is even expected that new quantum-chemical ideas will lead to new discoveries. To be ready for this event, our strenuous efforts are required. It will be helpful to establish quantum-chemical concepts as having distinct theoretical bases, and efforts in this direction should soon be fruitful.

It may even happen that an idea born as a result of such efforts, although it may be considered trifling by quantum chemists, will entirely change the historical flow of chemistry. What is more, such events will frequently occur from now on. Many jewels lie underground to be discovered by our efforts. If young quantum chemists with big dreams continue their research rigorously with confidence, outstanding results will surely be obtained. And so at this tune I would like to pass on my own dream, which I myself will never realize hereafter, to the young participants in this congress.

Some of the most important future aspects of chemistry will necessarily proceed in a nonempirical direction. The role of quantum chemistry in such development will of course become more and more important. I hope and expect that this congress will significantly contribute to the development of chemistry in this direction.

At the same time, we must attend to unexplored branches of chemistry, peripheral areas in particular. These can not be treated by analogy from experience and they are far beyond nonempirical rationalization. As a practical example, modern chemistry is faced with the great task of treating various problems of earth and mankind. In this connection, quantum chemical theories will produce new principles concerning the conversion of materials and energies. It is hoped that quantum chemists, not limited to this example, will become carefully but boldly concerned with the multifarious problems in the frontier fields. Quantum chemistry is expected to further strengthen the connection between various fields of science other than chemistry, thus deepening its interdisciplinary colour. Activity which is not confined to the frame of chemistry would be desirable. In this sense we expect the scope of discussions at this congress as well as the meaning of "quantum chemistry" used here will never be understood only in a limited or narrow sense.

On the basis of this idea, the prospects of quantum chemistry seem to be bright and broad, heading toward a vast expanse like an ocean of youthful energy. I believe not a few people agree with me that such a view is not too optimistic.

To all participants, particularly those from abroad, I wish to express a hearty welcome. We feel it our duty to exert ourselves during this congress to demonstrate to you the significance of your attendance, in return for your generous cooperation. May we all have a good and fruitful time.

Thank you for your attention!

WELCOME ADDRESS

K. Husimi, President of
Science Council of Japan

Mr. Chairman, Ladies and Gentlemen,

On the occasion of opening the Third International Congress of Quantum Chemistry I am honored with the chance of saying a few words on behalf of the sponsors.

First of all, I would like to extend my hearty welcome not only to Prof. Löwdin, President of the International Academy of Quantum Molecular Science and executive members of the Academy but also to all the participating scientists in this Congress who have come over to Japan from every continent. At the same time I cannot forget to express my heartfelt thanks to all the people concerned for their cooperation in arranging the Congress.

Chemistry is by nature an extensive discipline built up solely by a series of accumulated results from experimental laboratory work on chemical reaction. It belonged to a section of science characterized as most empirical and, nevertheless, gave birth to the modern atomic theory. When quantum mechanics appeared and developed to some extent, the foundation was firmly laid on which the nature of interactions and bindings of atoms and molecules could be accounted for solely on the theoretical considerations. The significance of quantum mechanics to chemistry is now accepted universally and provides a widely applicable methodology for attacking the problems of chemical phenomena.

Chemistry and physics which had originally been brought up differently from each other have become unified into the single discipline of molecular science. In the center of this science is placed quantum chemistry, which has more and more frequently has been applied to wider range of problems including those encountered in biology, pharmacy, medicine and engineering. Its development is really remarkable.

This Congress will deal with such themes as recently developed methods of quantum chemistry; newly cultivated sections of molecular fusion, fission and reciprocal action; remarkably expanded area of

K. Fukui and B. Pullman (eds.), Horizons of Quantum Chemistry, xi – xii.

molecular spectra; biological macro molecules and others. You, expert scientists, are engaged in the study of these themes in each country. I am sure it is really significant for you to get together for discussions of the latest research results.

The field of physics, in which I am personally interested, is rather different from this fascinating domain. So I am afraid I cannot comprehend fully the content of the scientific discussions in this Congress. However, I would like to add here one thing. Prof. Kotani, my senior in scientific field, has long been engaged in quantum chemical studies. I was privileged to watch him working on the extended calculations of molecular orbital integrals. In those days all numerical calculations had to rely on the manual calculating machine, which made me feel sorry for such a genius as Prof. Kotani wasting much of his precious time and talent in such barren work. Now that the development of electronic computers has made such remarkable progress, those with large mental capacities and abilities, such as Prof. Kotani, may use them positively in the scientific pursuits of quantum chemistry. May I add a personal account of mine that my doctorate thesis was on quantum mechanical density matrices? The same subject was later taken up and considerably extended by Prof. Löwdin. I imagine it may have helped him to start his study of quantum chemistry. I hope it can be accepted as a part of my qualifications permitting me to make this address.

On behalf of Science Council of Japan as a sponsor of this Congress I would like to say of its best wishes to you. Science Council of Japan has since its inauguration made every effort in performing its mission encouraging scientific researches and come to the conclusion that the scientific exchange among scientists of different nationalities is very important both in stimulating and evaluating each other at the scientific level.

All the participating scientists here are requested to make the most of this opportunity for strengthening international cooperation and gaining a variety of advantages.

International conferences have mostly been held in European countries and the U.S.A. This actually leads to a limited number of participants from Japan on account of the geographical inconveniences. Accordingly, Japanese scientists are less favored with the chance of presenting their research results to an international audience.

I would like the participating foreign scientists to take advantage of this opportunity for looking into the actual state of scientific research in Japan and at the same time for comprehending the life of Japanese people as well as the beauties of our traditional culture.

Last but not least, I sincerely hope the Congress will be successful and that the International Academy of Quantum Molecular Science will have a bright future.

Please accept my best wishes for a very profitable and enjoyable visit to Kyoto.

Thank you.

WELCOME ADDRESS

Yasuhide Yukawa, President of
The Chemical Society of Japan

Honorable President of the International Academy of Quantum Molecular Science, Guests, Ladies and Gentlemen,

As a member of the host organization, allow me to offer you my hearty congratulations on the opening of the Third International Congress of Quantum Chemistry in Kyoto at which many registrants are present.

Thanks to the great efforts of Professor Fukui, the chairman of the Organizing Committee, Professor Yonezawa, the secretary general, Professor Kotani, the chairman of the Fund Raising Committee, Professor Kodama, the co-chairman of the same Committee and all persons concerned, things are, fortunately, in fine trim for the Congress. Tomorrow, the presentation of about 350 papers and the discussions will begin. I expect that quantum chemistry and related sciences in Japan will be increasingly developed as a result of these papers and discussions. To the scientists who came to Japan for the Congress from many parts of the world, I sincerely hope that you will have the best understanding of Japan by getting a deeper understanding of quantum chemistry in Japan as well as the Japanese standard of overall science and technology.

Autumn is the best season of the year here when the old city of Kyoto is dressed up with red leaves. Here you will be able to observe various old traditions of Japan. You can visit the ancient city of Nara by taking 1-hour electric train ride from Kyoto. I sincerely hope that you will make new friends here and that you will enjoy autumn in Japan.

Thank you!

K. Fukui and B. Pullman (eds.), Horizons of Quantum Chemistry, xiii.

OPENING ADDRESS AT THE THIRD INTERNATIONAL CONGRESS OF QUANTUM CHEMISTRY

Per-Olov Löwdin, President of
The International Academy of Quantum Molecular Science

Mr. Chairman, Distinguished Colleagues, Ladies and Gentlemen,

On behalf of the International Academy of Quantum Molecular Science, it is my great privilege to welcome the participants in the Third International Congress of Quantum Chemistry arranged here in the beautiful and historic city of Kyoto, Japan. The initiative to these congresses is taken by the International Academy of Quantum Molecular Science, and the first congress was arranged in 1973 in Menton, France, and the second congress was held in 1976 in New Orleans, Louisiana, in the United States. We are very happy that the third congress is arranged here in Kyoto, and — on behalf of the Academy — I would like to thank Professor Kenichi Fukui and his Organizing Committee for their valuable work and tremendous efforts to organize a successful congress. I would also like to express our sincere gratitude to the Science Council of Japan and to the Chemical Society of Japan for their moral and financial support of the congress. Looking at the program, I am convinced that this congress, consisting of five symposia on selected fields with associated poster and oral sessions, will be a milestone in the development of quantum chemistry of great importance for the future.

Quantum chemistry is the application of the quantum theory of matter to the electronic structure of atoms, molecules, and solid-state investigating stationary states and spectra as well as chemical reactions and collision phenomena. The fundamental basis of quantum theory is the physical measurement, which implies an interaction between the object, the apparatus, and an observer, which can never be smaller than a "quantum of action". The measurement is complete when the information enters the mind of the observer through his perceptions, and — according to this philosophy — the physical universe is experienced as a projection on the mind of the observer — an idea which has deep roots in many of the ancient schools of thinking in the Orient.

In chemistry, matter consists of molecules, molecules of atoms, and atoms of a few elementary particles: electrons and atomic nuclei —

K. Fukui and B. Pullman (eds.), Horizons of Quantum Chemistry, xv–xviii.

the latter consisting of protons and neutrons — which do not obey the laws of classical physics but the rules of modern quantum theory formulated around 1925. The non-relativistic quantum chemistry has a sound axiomatic basis connected with the mathematical theory of the Hilbert space, and it is little influenced by all the complications of the high-energy physics with its divergence difficulties and menagery of "elementary particles" — even if these problems may have to be considered in the future.

In the natural sciences, one has always tried to explain all the many different phenomena in nature by means of a simple unifying principle. The theory of the electronic structure of matter based on the Schrödinger equation and its generalizations is probably the greatest unifying principle one has found so far. Today one knows millions of molecules built up from about one hundred of different atoms, which in turn are built up by identical electrons forming intricate wave patterns around the atomic nuclei, which — in this connection — may be considered as positive point charges characterized by the atomic numbers. In starting from these atomic numbers describing a molecular system, modern quantum chemistry has an ab-initio character which is otherwise rare in the natural sciences.

Even if, in axiomatic quantum theory, all physical phenomena are finally related to the "mind" of an observer, there is also another side to the story. The observer as a human being is undoubtedly a phenomenon in biology — the observer belongs to Homo sapiens. The unifying principle in biology is cellular biology, which is further reduced to submolecular biology dealing with the motions of electrons, protons, and molecular fragments. Professor Albert Szent-Gyorgyi has described life as a drama played by the electrons as actors on a stage formed by the biomolecules — only that the actors are one billiard (10^{-15}) times smaller than those we are accustomed to from an ordinary theatre. The theory of the electronic structure of life is treated in a special field called quantum biology.

Another way to express this unifying principle is to say that biology may be reduced to biochemistry, biochemistry to chemistry, chemistry to physics, physics to the quantum theory of the elementary particles, and quantum theory to the mind of the observer. Since the observer is part of biology, one is more or less back to the starting point — but one is not making a cycle but instead a "spirale" or helix providing a deeper and deeper knowledge for each turn.

If one reduces the quantum theory of matter to the mind of an observer, one has certainly an idealistic approach to science, whereas — if one reduces the observer to matter built up by elementary particles — one has a materialistic description. The two descriptions supplement each other, and quantum theory throws hence new light on the classical philosophical problem of the connection between mind and matter. The solution of this problem is of great practical importance in treating, for instance, mental diseases — and today one has good reason to believe that most (if not all) mental diseases are related to molecular and submolecular disturbances, and that they may be treated on this level.

In relation to the problem of the connection between mind and matter, I may perhaps tell a small joke about an English philosopher, who had found an excellent solution to this question: "What is mind — it does not matter. What is matter — never mind". Unfortunately, it is untranslatable.

In connection with quantum biology, I would like to mention that today there are many scientists who firmly believe that the origin of cancer is related to disturbances of the electronic structure of some of the fundamental biomolecules — particularly in the nucleoproteins, i.e. deoxyribonucleic acids in protein overcoats which form the genes, and in their control mechanisms. Some of these problems will be discussed at this congress and in the post — symposium on biomolecules also here in Kyoto.

However, there are also other important applications of the quantum theory of the electronic structure of matter and, in solid-state technology, one has the field of "quantum electronics", which includes the construction of transistors, masers and lasers, integrated circuits, and many new electronic "gadgets" under development. The basic research in this area forms the basis for the future development of the electronic industry — of particular importance here in Japan.

On the chemical side, there are many important applications connected with the theory of the chemical reactions, and the proper understanding of the phenomenon of catalysis and the construction of specific catalytic agents is of fundamental importance for the entire chemical industry. In this connection, one needs a quantum theory which contains time, temperature, and other macro-scopic parameters. Instead of the Schrödinger equation dealing with wave functions, one has to consider the Liouville equation for the system operators on density matrices, which describe the more complicated systems occurring in ordinary chemistry. One enters the field of "quantum statistics" dealing also with such concepts as entropy and free energy, of which "quantum mechanics" is only a special ideal case that may occur only at the absolute zero of temperature ($T = 0^oK$). In this field, one has also to study the irreversible phenomena which occur in nature, and which are of enormous practical importance in connection with the solution of the energy problem for our world.

Even if quantum chemistry today is slightly more than 50 years old, it is still at a "baby stage", where most of the really important problems remain to be solved — both as to fundamental theory and to applications. This ought to be a great challenge to the next generation of scientists, and particularly to the young scientists from Japan and overseas who are attending this congress. Your work in the future will be essential not only for the deepening of our basic knowledge but also for the technological and industrial applications of fundamental importance for the world economy and the standard of living in all countries.

This congress is an example of what can be achieved through international cooperation on a global basis, and we are all looking forward to a valuable exchange of scientific ideas. In this work, we are getting in contact with scientists from many different countries, and I am personally convinced that this will contribute also to a better understanding between the people of the world — in this way contributing also to peaceful coexistence, valuable cooperation, and global unification.

We are happy to be here in Kyoto, and we participants in the third international congress of quantum chemistry would like to thank our Japanese hosts for their warm welcome and for the immense hospitality they have already shown us. I am sure that the Kyoto Congress is going to be a wonderful scientific event which we will treasure in our memory for ever.

SYMPOSIUM I. NEW APPROACHES IN QUANTUM CHEMICAL METHODOLOGY

Chairman : M. Kotani

Science University of Tokyo
Tokyo, Japan

INTRODUCTION TO SYMPOSIUM I

Masao Kotani

The Symposium I of this morning will be the first scientific session of this Congress, and I have been given the privilege to be the chairman of this important Symposium.

During the half century since 1929, quantum molecular science has made a really remarkable progress. Since perhaps I am one of the oldest participants here, please allow me to return to 50 years ago, when I was graduated at the University of Tokyo. Quantum molecular science was then in a very primitive stage, being only 2 years after Heitler-London's theory of the hydrogen molecule opened this field. In 1929 Lennard-Jones' paper suggesting molecular orbitals was published, and basic papers by Profs. Mulliken and Hund appeared around this period. Slater determinant constructed of spin-orbitals followed quickly. Idea of density matrix was presented by Dirac and extended by Husimi, but not much attention to this concept was paid by molecular scientists in those days. After Roothaan found the standard formalism of the Hartree-Fock treatment of molecules, theoreticians' concern became focussed on correlation energy. Historically I think it interesting to note that Heitler-London's theory already included a part of the correlation energy, using 2 orbitals for a 2 electron system. I remember in this connection that Prof. Lowdin, present President of the International Academy sponsoring this Congress, nicely explained a major part of the correlation energy of 6 pi-electron system of the benzene molecule by using 6 alternant orbitals, at a Congress held in this city Kyoto in 1953.

In the last quarter century 1955-1979 rapid development of electronic computer provided us with more and more powerful means of calculation, and configuration interaction became a standard method to surpass Hartree-Fock approximation. Nowadays we hear that thousands of configurations can be incorporated in actual calculations. Quantum chemistry has been established as a discipline, which attracted not only younger chemists but also physicists and even mathematicians. Another important characteristic feature of this recent quarter century consists in many new theoretical developments in dealing with so-called many-body problems in quantum mechanics, some of which are shared with nuclear and solid-state physics. This situation is very encouraging for us, renewing and deepening our understanding of the nature of many-electron system in molecules, and showing a new horizon of quantum chemistry. Remarkable possibilities hidden in reduced density matrices were steadily sought for throughout this period.

The present Symposium is dedicated to lectures and discussion concerning new developments in such theoretical concepts and methods. We are favored by the presence of three invited lecturers: Profs. Robert Parr, Wilfried Meyer, Osvaldo Goscinski. I asked further Prof. F. A. Matsen to give a shorter communication as the fourth lecturer.

K. Fukui and B. Pullman (eds.), Horizons of Quantum Chemistry, 3.

DENSITY FUNCTIONAL THEORY OF ATOMS AND MOLECULES

Robert G. Parr

Department of Chemistry
University of North Carolina
Chapel Hill, North Carolina 27514, USA

ABSTRACT

Current studies in density functional theory and density matrix functional theory are reviewed, with special attention to the possible applications within chemistry. Topics discussed include the concept of electronegativity, the concept of an atom in a molecule, calculation of electronegativities from the Xα method, the concept of pressure, Gibbs-Duhem equation, Maxwell relations, stability conditions, and local density functional theory.

PREFACE

In this talk I shall try to reveal the scope, power and promise for chemistry of density functional theory. Necessarily I must be sketchy and incomplete. I will emphasize unpublished work from my laboratory on the last five of the following topics though I will start with the first:

Density functional theory
The concept of electronegativity
The concept of an atom in a molecule
Electronegativities from the Xα method
The concept of pressure
Gibbs-Duhem equation
Maxwell relations and stability conditions
Local density functional theory

My collaborators are the persons indicated below, all at the University of North Carolina except Norman March, with whom I completed a study in Oxford last month. I also may mention Peter Politzer, who did related work while visiting me at Johns Hopkins University. Professor Politzer, Professor Ray, Dr. Gadre and Dr. Nakatsuji are here at this meeting.

K. Fukui and B. Pullman (eds.), Horizons of Quantum Chemistry, 5–15.

Libero Bartolotti	George Henderson	William Palke
Robert Donnelly	Mel Levy	Philip Payne
Carol Frishberg	Norman March	Naba Ray
José Gázquez	Danny Murphy	Leonard Samuels
Shridhar Gadre	Hiroshi Nakatsuji	Wen-Ping Wang
Nicholas Handy		

DENSITY FUNCTIONAL THEORY

An atomic or molecular system of interest is defined by the number of electrons N and the external or nuclear-electron potential $v(\vec{r})$. Conventionally, one solves the Schrödinger equation to determine the 4N dimensional wavefunction. Equivalently, one solves the variational principle

$$\delta\{\langle\psi|\hat{H}|\psi\rangle - E\langle\psi|\psi\rangle\} = 0 \ . \tag{1}$$

That is, one minimizes the quantity $\langle\psi|\hat{H}|\psi\rangle$ subject to a normalization constraint, the Lagrange multiplier being the energy E. All observable quantities are functionals of ψ, readily computed once ψ is known. This includes the electronic kinetic energy T, the nuclear -electron potential energy V_{ne}, the electron-electron repulsion energy V_{ee}, and the 3 dimensional electron density

$$\rho\,(1) = N\int\cdots\int|\psi|^2 ds_1 dv_2\cdots dv_N, \tag{2}$$

$$N = \int\rho(1)\,d\tau_1 \ . \tag{3}$$

The procedure is a procedure to go from N and v to the energy and other properties. We may say that E is a functional of them,

$$E = E[N,v] \ . \tag{4}$$

Density functional theory [1] permits one to work with ρ in place of N and v. Unfortunately it is restricted in the first instance to ground states. But for them, ρ determines v (and of course N) and hence (through the conventional route) all properties. That is Kohn-Hohenberg Theorem I. There also is a variational principle, Theorem II. The correct density and energy may be found by minimizing the functional

$$E[\rho] = T[\rho] + V_{ne}[\rho] + V_{ee}[\rho], \tag{5}$$

keeping the number of particles fixed at the correct number for the system of interest. That is,

$$\delta\{E[\rho] - \mu N[\rho]\} = 0, \tag{6}$$

where μ is a Lagrange multiplier, called the chemical potential for the system.

One comment on Theorem II. It has recently been shown by Mel Levy [2] that the trial ρ need not be restricted to those ρ which are associated with ground states with local potentials.

The energy functional is the sum of three functionals, the known functional V_{ne},

$$V_{ne}[\rho] = \int v\rho d\tau, \tag{7}$$

and the two unknown functionals T and V_{ee}. The functional V_{ee} itself is conveniently expressed as the sum of two parts, a known classical coulomb energy J of the charge distribution,

$$J[\rho] = \frac{1}{2}\int\int \frac{\rho(1)\,\rho(2)}{r_{12}}\, d\tau_1 d\tau_2, \tag{8}$$

and an unknown exhange-correlation energy $-K$. If we introduce a new functional F, the sum of T and V_{ee}, then the theory can be expressed in terms of it. The stationary principle is

$$\mu = v + \frac{\delta F}{\delta \rho} = \text{constant.} \tag{9}$$

One must find the ρ which makes the indicated quantity constant through space (and E a minimum).

Levy [2] has given a prescription for finding the number $F[\rho]$ from a given ρ. Compute the integral $\langle\psi|\hat{T} + \hat{V}_{ee}|\psi\rangle$ for each and every antisymmetric wavefunction ψ which has the given density ρ. The minimum value of all these numbers is the desired $F[\rho]$.

THE CONCEPT OF ELECTRONEGATIVITY

The chemical potential μ is just the electronegativity of chemistry, up to a sign [3]. To see this, note first that μ is the derivative $(\partial E/\partial N)_v$, which follows from Eq.(9). Then note that

that is a slope of an E versus N curve at constant v. And then remember finally that this slope already has been identified with the electronegativity χ by many authors, notably Mulliken, whose classic formula $\chi = 1/2\ (I + A)$ is the finite difference (parabolic) approximation to the (negative) slope of the E versus N curve for an atom, computed from its ionization potential and electron affinity.

All the properties of electronegativity follow [3]. We may summarize:

(1) Electronegativity is a characteristic property of a ground state of an atom or molecule, equal to the negative of the chemical potential of density functional theory.

(2) Electrons tend to flow from a region of low electronegativity to a region of high electronegativity.

(3) When atoms come together to form a molecule, differing electronegativities become equal by charge transfer (Sanderson's Principle of Electronegativity Equalization [4]).

(4) In an exact orbital description of an atom or molecule, every orbital has the same electronegativity.

Property (4) requires for its proof an extension of density functional theory to first-order density-matrix functional theory, which has been carried out by Gilbert [5] and Donnelly [6], and which I do not have time to here discuss. Neither do I have time to describe the quantitative studies we have already carried out regarding electronegativity equalization [7].

GIBBS-DUHEM EQUATION

It is important to know how the electronegativity changes from one circumstance to another. The energy change itself is given by the formula

$$\delta E = \mu \delta N + \int \rho \delta v d\tau, \tag{10}$$

which you will recognize as the Hellmann-Feynman formula generalized to include change of number of particles. To get the chemical potential change, insert Eq.(9) into Eq.(5), take the differential, and equate to Eq.(10). The result is

$$N \delta \mu = \int \rho \delta v d\tau + \delta Q, \tag{11}$$

where the universal functional Q is defined by the formula

$$Q[\rho] = -F[\rho] + \int \frac{\delta F}{\delta \rho}\, \rho\, d\tau. \tag{12}$$

Equation (11) is analogous to the Gibbs-Duhem equation of macroscopic thermodynamics.

THE CONCEPT OF AN ATOM IN A MOLECULE

I now give a density functional answer to the question: What is an atom in a molecule? The argument is a simplification of the one already published [3].

Consider a molecule AB in its ground state, built from atoms A and B in their ground states. All these entities have characteristic ρ, N, v and μ. In the molecule A and B are in states A^*, B^*, possessing densities ρ_A^*, ρ_B^*, numbers of electrons N_A^*, N_B^*, chemical potentials μ_A^*, μ_B^*, in perturbed environments v_A^*, v_B^*. We require

$$\rho_A^* + \rho_B^* = \rho_{AB}, \tag{13}$$

$$\mu_A^* = \mu_B^* = \mu_{AB}, \tag{14}$$

and minimize the energy change

$$\Delta E^* = \{E_A[\rho_A^*] - E_A[\rho_A]\} + \{E_B[\rho_B^*] - E_B[\rho_B]\} . \tag{15}$$

The atoms A^* and B^* are thereby completely defined! Furthermore, to first order the charge transferred is proportional to the initial electronegativity difference $\mu_A - \mu_B$, and the error in energy if charge transfer is ignored is proportional to the square of this difference.

DIFFERENTIAL COEFFICIENTS AND STABILITY CONDITIONS

Second-order effects of course are very important; we have just begun to analyze them. In effect one has thermodynamics for a stable (ground) state, so that we may use thermodynamic methods. The fact that δE is an exact differential leads to formulas for the differential coefficients and Maxwell relations; the fact that the state is stable leads to inequalities. There are many interesting applications of the resulting formulas. For example, there is the formula for an atom

$$Z(\partial\mu/\partial Z)_N = (\partial V_{ne}/\partial N)_Z . \tag{16}$$

Or there is the fact that the slope of the E versus N plot for atoms monotomically increases with N.

ELECTRONEGATIVITIES FROM THE Xα METHOD

I now turn to the calculation of electronegativities by one technique, the Xα transition state method [8]. I believe that this technique is best regarded as an approximation to the exact first-order density-matrix functional theory of Gilbert [5] and Donnelly [6], and so I first briefly describe that.

Since ρ determines everything and the first-order density matrix γ determines ρ, γ determines everything. Since $T[\gamma]$ is known, all that one needs to know to proceed with implementation is the exchange-correlation functional $K[\gamma] = K[\rho]$. The exact equations for the chemical potential (electronegativity) are

$$\mu = (\partial E/\partial n_i)_{n_j \chi}, \tag{17}$$

where the n_i are the occupation numbers for the natural spinorbitals χ. In the simplest (spin nonpolarized) Xα method (which we use), one makes the single assumption that K is local. This is enough to assure, by an analysis of Szasz and Berrios-Pagan [9], that K has the form

$$K[\rho] = \frac{3}{4} C\alpha \int \rho^{4/3} d\tau, \tag{18}$$

where α depends on N, and that is the Xα method. Equal electronegativities for all orbitals are lost, but in each particular case one may proceed to compute the derivative of the total energy with respect to total number of electrons, along the ground state path. More specifically, in accord with our previous discussion we take the chemical potential to be half the energy difference on going from positive ion to negative ion, computed by the Xα transition state method.

The first 54 atoms in the periodic table are covered by just two cases. Case A is the case in which only one orbital is involved in going from positive ion to negative ion --- the typical open shell atom. Then the electrogegativity is the negative of the orbital energy for this orbital in the transition state, which usually but not always is the ground state of the neutral atom. Case B is the case in which two orbitals are involved in the transition between positive ion and negative ion --- the typical closed shell atom. Usually, but not always, the transition state is identical with the transition state for the excitation of the neutral atom to its first excited state. The electronegativity is the negative of the average of the corresponding two orbital energies.

Numerical values of electronegativities obtained by this scheme are excellent [8], especially if we view it as just the first approximation to an exact method, an approximation which admits to systematic improvement. One notes in the numbers what will also be seen in other tables of electronegativities and is well known. Electronegativity decreases slowly as one goes down a column in the periodic table.

SCALING ARGUMENT FOR NEUTRAL ATOMS

I pass now to a discussion of the trend just mentioned in electronegativity values. The result will be the asymptotic law [10]

$$\mu \propto Z^{-1/3}. \tag{19}$$

In passing I will remark on a relation of Politzer and Ruedenberg that recently has received considerable attention [11].

It is well known that the so-called Thomas-Fermi theory of neutral atoms gives two results, $\mu = 0$ and $E = -3V_{ee}$ [3]. It also gives the Politzer-Ruedenberg relation, $E = (3/2)\Sigma\varepsilon_i$.

To get a Z-dependent μ, a systematic procedure is available. By rearranging a formula from 1/Z expansion theory, one gets the expansion

$$E = Z^{7/3}f_1(N/Z) + Z^{6/3}f_2(N/Z) + Z^{5/3}f_3(N/Z) + \cdots . \tag{20}$$

If one then invokes the Lieb-Simon theorem that Thomas-Fermi theory becomes exact as Z approaches ∞, one sees that the derivatives of the first five f_k, evaluated at $N = Z$, are zero, and the first nonvanishing contribution to μ comes from the sixth f_k, and gives the behavior of Eq.(19).

A corresponding ansatz can be written down for molecules [10]. Again the first term corresponds to Thomas-Fermi thoery and gives the Politzer-Ruedenberg relation; the sixth and later terms describe the chemical potential.

THE CONCEPT OF PRESSURE

Back to formal density functional theory, I now develop the concept of local pressure in an atom or molecule [12].

The stationary formula for μ, Eq.(9), may be rewritten in terms of the total classical electrostatic potential v^* and the functional derivative of a new universal functional $G = T - K$:

$$\mu = v^* + \frac{\delta G}{\delta \rho} = \text{constant.} \tag{21}$$

In the absence of v^*, the chemical potential would be just $\delta G/\delta\rho$; call this μ°. Then Eq.(21) is analogous to the equation for a chemical potential in a gravitational field, $\mu = \mu^\circ + mgh$, and we are prompted to make definitions of a stress tensor $\overleftrightarrow{\sigma}$ and the scalar pressure P as follows:

$$\overleftrightarrow{\sigma} = \sum_{i=1}^{3} \sum_{j=1}^{3} \vec{\varepsilon}_i \sigma_{ij} \vec{\varepsilon}_j , \tag{22}$$

$$\vec{\nabla} \cdot \overleftrightarrow{\sigma} = -\rho \vec{\nabla}[\delta G/\delta\rho], \tag{23}$$

$$P = -\frac{1}{3} \text{trace } (\sigma_{ij}) . \tag{24}$$

Among other interesting results, one then finds

$$3\int P d\tau = 2T[\rho] - K[\rho] . \tag{25}$$

Also

$$E = \int \rho \mu d\tau - \frac{1}{2}\int \vec{\varepsilon}\cdot\vec{\varepsilon}\, d\tau - \int P d\tau - X[\rho] \tag{26}$$

and

$$N\delta\mu = \int \rho\, \delta v^* d\tau + \int \delta P d\tau + \delta X, \tag{27}$$

where the quantity X is defined by

$$X[\rho] = \left[\int \rho \frac{\delta T}{\delta \rho} d\tau - \frac{5}{3} T\right] - \left[\int \rho \frac{\delta K}{\delta \rho} d\tau - \frac{4}{3} K\right]. \tag{28}$$

These formulas are extremely interesting in the way they expose the essential importance of the universal functional $X[\rho]$. If $T[\rho]$ were a local functional, that is, an integral of a local function of ρ, that function would be the five-thirds power of ρ [9], and the contribution of T to X would vanish. Similarly, if $K[\rho]$ were an integral of a local function of ρ, that would be the four-thirds power, and the contribution of K to X would vanish. $X[\rho]$ therefore measures the nonlocality of $T[\rho]$ and $K[\rho]$.

The functional $X[\rho]$ vanishes in Thomas-Fermi theory and in Thomas-Fermi-Dirac theory. It also vanishes in a theory I shall now describe, which we call local density functional theory.

LOCAL DENSITY FUNCTIONAL THEORY

In this theory [13] we make a local approximation on the whole $V_{ee}[\rho]$, not on the exchange-correlation part of it, and we make the usual local approximation to $T[\rho]$. That is, we work with the functional

$$E_L[\rho] = \frac{3}{5} A(N)\int \rho^{5/3} d\tau + \frac{3}{4} B(N)\int \rho^{4/3} d\tau + \int v\rho d\tau , \tag{29}$$

where the coefficients A(N) and B(N) are to be determined. The results are extremely simple; they provide an alternative to Thomas-Fermi-Dirac theory which has many advantages over it.

A single assumption suffices to determine A(N) and B(N): $E = -0.6127\ Z^{7/3}$ for neutral atoms. One then finds $\mu = 0$, $B(N) = 1.0058\ N^{2/3}$ and $A(N) = 6.4563$.

For atoms, simple closed analytical formulas result for everything. For 625 atoms and ions, the energy is fit to a root-mean-square error of 5%, which is a better fit than given by any simple theory we know of. Atoms have finite sizes and negative ions can be stable.

For molecules, the solution again is very simple. The density is an algebraic function of the bare-nuclear potential; contours of v are contours of ρ. This prompts me to ask: How close is this to being true for accurate molecular wavefunctions? Perhaps some one of you will carefully investigate this question, the answer to which surely is important to know.

Probably it seems strange that V_{ee} is even approximately representable as the integral of a power of ρ times $N^{2/3}$. In fact, however, V_{ee} can quite accurately be so represented. One reason is that there exists a fairly tight bound for any density [14],

$$J[\rho] \leq 1.092\ N^{2/3} \int \rho^{4/3} d\tau \ . \tag{30}$$

This is the principal part of V_{ee}, and the actual value of J for Hartree-Fock atomic densities is amazingly close to this bound.

To extend the argument, it would be sensible to approximate the whole V_{ee} for atoms and molecules with a more general function of N times the integral of $\rho^{4/3}$. If, indeed, we take the simple form

$$V_{ee} = (B_0 N^{2/3} + B_1) \int \rho^{4/3} d\tau, \tag{31}$$

we find a strikingly good fit of Hartre-Fock values of V_{ee} for atoms [14]. The parameter B_1 is negative, and its value agrees well with the Dirac formula for exchange energy.

Such a local theory constitutes a natural starting point for systematic development of more accurate theory, for example by what are called gradient expansion techniques. We have done considerable work with such techniques [15], but time does not allow me to go into them here.

CONCLUSION

To conclude, here is a list of some of the problems on which we are currently working:

- Calculation of μ by spin polarized Xα method
- Calculation of μ by MCSCF-type method
- Is μ a minimum for the ground state?
- Detailed elucidation of second-order effects
- Improved description of $T[\rho]$
- Improved description of $K[\rho]$ and/or $K[\gamma]$
- Gradient expansion corrections to local density functional theory
- Corresponding time-dependent theory
- Corresponding theory for excited states

Most these problems are difficult, and substantial progress is assured on only a few of them. But I believe, nevertheless, that our knowledge of density functional theory will continue to expand over the years ahead, and that density functional theory will contribute substantially to the ultimate quantum-theoretical elucidation of chemistry.

I thank Professor Keiji Morokuma of the Institute for Molecular Science, Okazaki, Professor Jiro Tanaka of Nagoya University, Nagoya, and the Japan Society for the Promotion of Science, for enabling me to spend October and November, 1979, in Japan, where this text was prepared.

REFERENCES

1. Hohenberg, P. and Kohn, W.: 1964, Phys. Rev. B 136, pp.864-871.
2. Levy, M.: 1979, Proc. Natl. Acad. Sci. USA, in press.
3. Parr, R. G., Donnelly, R. A., Levy, M., and Palke, W. E.: 1978, J. Chem. Phys. 68, pp.3801-3807.
4. Sanderson, R. T.: 1955, Science 121, pp.207-208.
5. Gilbert, T. L.: 1975, Phys. Rev. B 12, pp.2111-2120.
6. Donnelly, R. A. and Parr, R. G.: 1978, J. Chem. Phys. 69, pp.4431-4439; Donnelly, R. A.: 1979, J. Chem. Phys. 71, pp.2874-2879.
7. Ray, N. K. Samuels, L., and Parr, R. G.: 1979, J. Chem. Phys. 70, pp.3680-3684.
8. Bartolotti, L. J., Gadre, S. R., and Parr, R. G.: 1979, J. Am. Chem. Soc., submitted.
9. Szasz, L. and Berrios-Pagan, I.: 1975, Z. Naturforsch. 30a, pp.1516-1534.
10. March, N. H. and Parr, R. G.: 1979, J. Chem. Phys., submitted.
11. Politzer, P. and Parr, R. G.: 1974, J. Chem. Phys. 61, pp.4258-4262; Politzer, P.: 1976, J. Chem. Phys. 64, pp.4239-4240; Ruedenberg, K.: 1977, J. Chem. Phys. 66, pp.375-376; Parr, R. G. and Gadre, S. R.: 1979, J. Chem. Phys. submitted.
12. Bartolotti, L. J. and Parr, R. G.: 1979, J. Chem. Phys., in press.
13. Parr, R. G., Gadre, S. R., and Bartolotti, L. J.: 1979, Proc. Natl. Acad. Sci., USA 76, pp.2522-2526.
14. Gadre, S. R., Bartolotti, L. J., and Handy, N. C.: 1979, J. Chem. Phys., in press.
15. Wang, W-P., Parr, R. G., Murphy, D. R., and Henderson, G. A.: 1976, Chem. Phys. Letters 43, pp.409-412; Murphy, D. R. and Parr, R. G.: 1979, Chem. Phys. Letters 60, pp.377-379 (1979); Murphy, D. R. and Wang, W-P.: 1979, J. Chem. Phys. in press; Shih, C. C., Murphy, D. R., and Wang. W-P.: 1979, Chem. Phys. Letters, submitted.

SOME RECENT ADVANCES IN THE USE OF PROPAGATOR METHODS IN QUANTUM CHEMISTRY. FROM AMO TO AGP

Osvaldo Goscinski
Department of Quantum Chemistry
Uppsala University, Box 518
S-751 20 Uppsala, Sweden

ABSTRACT

A survey of recent developments in the use of propagator methods in quantum chemistry is made. Particular emphasis is devoted to balanced approximations of both states and operators in order to get a consistent description of both static and dynamic properties. A systematic procedure provides, in particular, an alternative derivation of a significant result by Linderberg and Öhrn: the antisymmetrized geminal power wave function (AGP): $\Psi_{AGP} = NAg(1,2)g(3,4)\ldots g(N-1,N)$ is the correct ground state of the self-consistent particle-hole propagator (SCPHP) (N is a normalization constant, A an antisymmetrizer and g an antisymmetric, normalized geminal). This AGP function was previously dismissed as a reasonable approximation on account of the limited amount of correlation energy it seemed to yield or because its inability to lead to proper dissociation. It is shown that on the contrary, provided no restrictions on the geminal are made, it contains as particular cases the different orbitals for different spins (DODS) scheme and the alternant molecular orbital method (AMO) of Löwdin - both capable of high accuracy. It is argued that the current theories of chemical reactions can have the AGP function as an optimal framework, with AMO as the simplest level of formalization of the methods of Bader, Fukui, Pearson, Woodward and Hoffman. It encompasses also the more advanced level proposed by Ruedenberg in terms of orbital reaction space and natural reaction orbitals. Some consequences of the AGP function being the ground state of the SCPHP are: the excitation spectrum is easily obtainable; the excited states are obtained by "one-electron excitation operators"; the validity of the Hellman-Feynman theorem which leads to a simple description of the response to external perturbations.

Essential to the previous aspects is the non-singlet character of the geminal. This "broken symmetry" allows the inclusion of a major part of the correlation energy. At the orbital level and the Hartree-Fock (HF) self-consistent (SCF) level of approximation these low-symmetry orbitals provide reasonable transition energies and probabilities in core-hole ionization as well as in $n-\pi^*$ transitions through the use of orbitals

K. Fukui and B. Pullman (eds.), Horizons of Quantum Chemistry, 17–35.

adapted to the excitation process: transition orbitals.

INTRODUCTION

In conventional quantum chemistry state properties, excitation energies and transition probabilities are usually described with approximate wave functions. Operators are replaced by their matrix representatives in a Hilbert space, in the spirit of Schrödinger. Time evolution and the Heisenberg "picture" would require to approximate the operators in some manner that does not violate the equations of motion. The conventional procedure accurate as it might be does not guarantee that sum-rules and operator algebra are not violated. The direct calculation of transition energies and probabilities by propagator or Green's function methods involves bypassing the wave functions /1/. It has been advocated in atomic and molecular problems by Linderberg and Öhrn /2/ and has been quite successful /3,4/ as well as the work of Cederbaum et al. /5,6/ on photoelectron spectroscopy and Öhrn and collaborators /7/ on electron binding energies. The related approaches based on equations of motion (EOM) by Simons, McKoy, Freed and others have been equally accurate /8-10/. It is not the purpose of the present lecture to describe these methods in detail but rather to discuss some particular developments which establish contact between wave function and operator methods. The generalization to ensembles is immediate in propagator methods. This is not true for wave function ones. Wave function methods have changed a great deal since Coulson said during his 1953 Japan visit: ..."I believe that the most urgent need in this field is to discover by one means or another which configurations we must include and which we may leave out, if we are not brave enough to include them all"... /11/. Calculations using the direct configuration interaction method (CI) /12/ and super CI /13,14/ are quite brave! In direct CI 10^5 selected configurations are nowadays possible. The developments using unitary groups and their associated diagrammatic techniques make these feasible /15/. But we are far perhaps from being brave enough if this means including, e.g., $\sim 10^{16}$ configurations (for S=0, 40 electrons and 40 orbitals).

A suitably chosen non-linear approximation may be a very forceful alternative to large-scale linear expansions. In 1953, also in Japan, Löwdin proposed in the spirit of the different bands for different spins treatment of antiferromagnetism by Slater /16/, a DODS treatment: the AMO method /17/. The basic idea was to account for the Coulomb hole introducing the semilocalized orbitals of Coulson and Fischer /18/. The method soon proved to be useful /19/ and it led to a rather successful account of the correlation energy (of the order of 90%) even in the restricted case when only one non-linear parameter is used /20/. In the present context it is interesting to point out that this approximation is perhaps "brave enough" in the sense of Coulson: all the determinants evenly excited from a reference configuration are included. The AMO function is essentially one determinant, projected to yield the right symmetries if necessary. It is an extension of the HF method which is of importance in the present context for the reasons discussed below, and well applicable beyond the realm of alternant systems.

The time-dependent Hartree-Fock (TDHF) method is a well established way of calculating atomic and molecular polarizabilities /2,3/. It is often referred to as the random-phase approximation (RPA) with exchange on account of its origin in the theory of extended isotropic systems /21/. In the usual derivations of the TDHF method one assumes the HF determinant to be the ground state and one then proceeds to construct excitation operators (leading to excitation energies, oscillator strengths, etc.) which no longer have the HF determinant as their reference state. This is a well-known feature /2,3/ in extended systems but it is particularly serious in the theory of finite systems. Is there a ground state associated to those operators? Is it a proper fermion state, appropriate for electrons? Is the number of electrons an integer? Is it the correct one? These questions, which are usually referred to as the N-representability problem /22,23/ are automatically solved when one uses wave function methods. To ignore them could lead to clear violations of representability /3/ and as such to make the physical validity of the results somewhat questionable. Linderberg and Öhrn showed that an AGP function is the appropriate ground state of the SCPHP /24,25/. They obtained this result and gave a prescription for constructing excitation energies and transition moments (in terms of poles and residues, respectively, of the propagator) using properties of the generators of the unitary group U(M) where M is the rank of the one-electron basis. This is related to the algebraic treatment by Fukutome /26/. In a series of papers we have approached the systematic approximation of propagators /27-33/. For the polarization propagator in particular we could obtain a result similar to the one by Linderberg and Öhrn /31/ and an alternative construction which proceeds by a previous optimization of the AGP function /32/. This leads to a practical procedure for calculations /33/. The AGP function is considered as a projected (insofar the number of electrons) Bardeen--Cooper-Schrieffer function (BCS) /34-36/. It has been very useful in the theory of superconductivity /37/ and superfluidity /38/. In quantum chemical context it has received consideration by Bratoz and coworkers /39-41/ but without yielding more than 40% of the correlation energy. Kutzelnigg analyzed the general problem /42/ and Watson's /43/ Be wave function in particular and observed that the AGP function does not approximate it well /44/. More recently, Linderberg verified that it does yield approximately 50% of the correlation energy in a model calculation /45/. Furthermore, it seems that it does not dissociate properly /45/. Nevertheless, simple considerations indicate that the AGP function, considered as a superposition of determinants, leads to the same excitation pattern as the AMO function. In this paper it will be shown in fact that an unrestricted geminal, i.e., with no special point group symmetry, with neither singlet nor triplet character, leads to an AGP function capable of dissociating correctly. The AMO scheme becomes a special case of the optimal AGP function. As such we can expect high accuracy from the AGP functions, with a pairing scheme which may be dominated by the AMO coupling, but not fully determined by it. We can conceive computational procedures related to the super CI methods /13-14/ but simplified by the rather simple nature of the second order reduced density matrix

/46/ as studied by Coleman /36/ and Rosina /47/. One can, furthermore, construct approximate propagators associated to the AMO method.

The AMO method emerges as a precursor of the AGP function. The combined analysis of states and operators leads to a richer and more stringent characterization of approximations and models. That the RPA has no consistent ground state has not been of major concern to many workers. This negligence, as pointed out by Linderberg /48/ has caused some reluctance to accept propagator methods in atomic and molecular physics. In a similar way elaborate wave functions not interrelated by meaningful excitation operators should perhaps become anathema.

EXCITATION OPERATORS AND PROPER GROUND STATES

Let $|0\rangle$ be the ground state representative of an electronic system. The excitation operators $\Omega_K^+ = |K\rangle\langle 0|$, for $K \neq 0$, verify the following properties:

$$\Omega_K^+|0\rangle = |K\rangle; \quad \langle K|L\rangle = \delta_{K\ell}; \quad \Omega_K|0\rangle = 0 \tag{1}$$

The last condition is an important property. Notice that the operators Ω_K^+ are many-electron operators. In the case of ionization an electron attachment one might be interested in specifying the number of electrons:

$$\Omega_K^+|N;0\rangle = |N\pm 1;K\rangle \tag{2}$$

which makes it natural to use the algebra of second quantization. An example of (2) is given by the Hartree-Fock (HF) method where $|0\rangle \sim |1,\ldots i,\ldots,N\rangle$, i.e., we have a single determinant of occupied spin orbitals $1 \leqslant i,j,\ldots \leqslant N$ (with virtual or "unoccupied" ones $N+1 \leqslant a,b,\ldots$). We have then approximations to the operators (2): $\{\Omega_K^+\} = \{a_1,\ldots,a_i,\ldots, a_N, a_{N+1}^+,\ldots, a_a^+,\ldots\}$. These fulfil (1). More explicitly:

$$a_i|N;0\rangle = |N-1;i\rangle; \quad a_a^+|N;0\rangle = |N+1,a\rangle, \tag{3}$$

i.e. the excited states are obtained by removing from electron in spin orbital i adding an electron in spin orbital a. This framework with a frozen orbital description is not adequate enough. It encompasses the Koopmans level of approximation /49/ but it neglects relaxation and correlation corrections /50-51/. To include them one needs operators of the form $\underline{h}_j$ in the classification of Pickup and Goscinski /28/:

$$\underline{h}_1 = \{a_i\}, \{a_a\}; \quad \underline{h}_3 = \{a_a^+ a_i a_j\}, \{a_i^+ a_a a_b\}. \tag{4}$$

These lead to a systematic procedure for obtaining corrections to Koopmans's approximation. It is clear though that for correlated wave functions (1) is not fulfilled with operators from $\underline{h}_1$. There is a delicate interplay between operator and state approximations. Concentrating in the following to excitations one could try to implement (1) with the particle-hole operators:

$$\Omega^+_{ai} = a^+_a a_i ; \quad \Omega_{ai} = a^+_i a_a \tag{5}$$

It is clear that the operators Ω^+_{ai} fulfil the conditions (1). They "excite" from the HF determinant to the singly replaced determinantal state $|1,\ldots,i-1,a,i+1,\ldots,N|$. But this is far from being a good description of excited states. The singly excited states interact with each other, with the doubly excited ones and hence indirectly with the ground state. (There is a wealth of experience accumulated regarding these matters /52/.) The operators in (5) are representative "one-electron operators" in second quantization. A natural extension to (5), and as means of approximating the many-electron operators of (1) is to devise a systematic procedure for generating excited states, fulfilling (1), with one-electron excitation operators:

$$\Omega^+_K = \sum_{i,a} \{a^+_a a_i Z_{ai,K} + a^+_i a_a Y_{ia,K}\}. \tag{6}$$

In TDHF one assumes operators of this form and also a single determinant ground state, which cannot possibly be the case. The method has been, nevertheless, very successful, in particular on account of the fulfilment of sum rules /53/. On the other hand, since the method leads also to a prescription for constructing the ground state energy, the question of what is the relevant wave function has been a puzzle for many years. This is specially true for large systems where our <u>ab initio</u> techniques are out of question, e.g., in the bonding of molecular or rare gas crystals /54,55/. In order to appreciate this fully we must consider the polarization propagator and its spectral resolution /1-4/.

PARTICLE-HOLE PROPAGATOR AND AGP

Consider the exact propagator $\langle\langle X^+,Y\rangle\rangle_E$, associated to a pair of operators X^+ and Y, which for our purposes may be defined by its spectral resolution:

$$\begin{aligned} \langle\langle X^+,Y\rangle\rangle_E &= -\langle 0|X^+T_o(-E+E_o)Y+YT_o(E+E_o)X^+|0\rangle \\ &= -\sum_{K\neq 0} \{\langle 0|X^+|K\rangle\langle K|Y|0\rangle/(-E+E_K+E_o)+\langle 0|Y|K\rangle\langle K|X^+|0\rangle/(E-E_K+E_o)\} \end{aligned} \tag{7}$$

This is no more mysterious than a generalized polarizability. It has poles for $E = \pm(E_K - E_o)$, i.e., for the excitation energies of the system and the residues are transition moments /2/. The reduced resolvents in (7) are defined by Löwdin as $T_o(\varepsilon) = P(\varepsilon - PHP)^{-1}P$ where $P = 1-|0\rangle\langle 0|$ /56/. In conventional treatemnts one approximates $|0\rangle$, E_o and the resolvent more or less separately and hopes for the best. Using the property $H|K\rangle = E_K|K\rangle$ one can see that (7) is equivalent to the following EOM /2/:

$$E\langle\langle X^+,Y\rangle\rangle_E = \langle 0|[X^+,Y]|0\rangle + \langle\langle X^+,[Y,H]\rangle\rangle_E \tag{8}$$

This leads to the approximation of "higher-order" propagators, as appearing in the r.h.s. of (8) in terms of "lower-order" ones, as appearing in the l.h.s. which is denoted by "decoupling". Decoupling schemes lead in turn to self-consistent procedures which bypass the explicit construction or knowledge of E_0 and/or $|0\rangle$. These are wrought with dangers of two kinds: the unsystematic nature may lead to severe symmetry problems or artifacts /56/ and the absence of $|0\rangle$ may imply lack of representability which might be disastrous (though not necessarily). A systematic procedure was introduced /28,29/, using superoperators and their inner projections. Let $\hat{1}$ and $\hat{H}$ be superoperators (the identity, $\hat{1}X=X$ and the Liouville-Kubo $\hat{H}$, $\hat{H}X=[X,H]$ for any operator X). Then defining

$$(X|Y)_0 = \langle 0|[X^+,Y]|0\rangle \text{ and } \hat{R}(E)X = (E\hat{1}-\hat{H})^{-1}X \tag{9}$$

one can write (8) in the following way /28,29/:

$$\langle\langle X^+,Y\rangle\rangle_E = (X|(E\hat{1}-\hat{H})^{-1}Y)_0 \tag{10}$$

Useful notation prompts useful approximations. We may now use the theory of inner projections /57/ for systematical approximation of resolvents. Notice though that one approximates superoperators in terms of operator matrix representatives instead of operator in terms of vector space matrix representations:

$$\langle\langle X^+;Y\rangle\rangle_E \sim (X|\underline{h})_0(\underline{h}|E\hat{1}-\hat{H}|\underline{h})_0^{-1}(\underline{h}|Y)_0 \tag{11}$$

for arbitrary choices of the operator manifold $\underline{h}$. In the case when we consider an array $\underline{X}$ of operators and limit $\underline{h}$ to coincide with $\underline{X}$, we have the <u>geometric</u> approximation /3,4/. These methods have been extensively applied /1,3,4/. Nevertheless, only the systematic construction of operators was touched, i.e., the nature of the ground state, if any, in the self-consistent approximations arises. These self-consistent procedures follow from contour integrations in (7) which lead to elements of the reduced second order density matrix /2-4/, since X and Y in (7) are particle-hole and hole-particle operators (of the form (5)). In order to construct the inverse matrix in (11) one needs both the first and second order reduced density matrices. Not all the information is simply available and the SCPHP did not lead to representable density matrices /3/. Consideration of this problem and, in particular, of the way to make the self-consistent procedures as consequence of the theory rather than an extra condition makes the Gel'fand-Naimark-Segal (GNS) construction a natural tool for performing propagator approximations /30/. The consistent use of this formulation makes seemingly unrelated procedures fall into the same class.

The actual solution for the particle-hole propagator in its self-consistent form was given by Linderberg and Öhrn /24,25/. The ground state has AGP form. An energy optimized AGP, leading to E_{AGP}, can be used to generate a manifold of excited states:

$$|K\rangle = O_K^+|AGP\rangle; \quad O_K|AGP\rangle = 0 \tag{12}$$

where the excitation operators O_K^+ are combinations of particle-hole and hole-particle operators in a Brueckner spin orbital basis /31/. The SCPHP can be expressed in the form (7), but with $|0\rangle$ replaced by $|AGP\rangle$, E_0 by E_{AGP} and P spanned by the projectors $|K\rangle\langle K|$. This result relates hermiticity conditions, stationarity, completeness and non-redundance of the operator manifolds previously discussed by various authors /26, 1,58,59/. These questions are not entirely trivial and one had to develop concepts which replace the conventional notions of approximation of functions. Our derivation of (12) proceeded not by the properties of the unitary group U(M) but by the implementation of some general techniques from field theory, and using the GNS construction. The outcome can be formulated quite simply: consider a normalized AGP function (for simplicity we consider here the case when N is even):

$$\Psi_{AGP} \equiv g^{N/2} = CAg(1,2)g(3,4)\ldots g(N-1,N), \tag{13}$$

with g in its canonical form in terms of 2x2 normalized determinants of natural spin orbitals u_i

$$g(1,2) = \sum_{i=1}^{M/2} \xi_i |u_{2i-1}(1)u_{2i}(2)|; \quad \sum_{i=1}^{M/2} |\xi_i|^2 = 1. \tag{14}$$

If it has been obtained by making the energy functional

$$E[\Psi_{AGP}] = \langle g^{N/2}|H|g^{N/2}\rangle / \langle g^{N/2}|g^{N/2}\rangle$$

stationary with respect to variations in the $\{\lambda_i\}$ as well as a local minimum with respect to variations of the 1-particle states, it is a consistent ground state for the particle-hole propagator /31-32/.

This, per se, would have only academic interest if the minimization of (8) would be difficult. But the second order reduced density matrix $\underline{D}^{(2)}(g^{N/2})$ associated to Ψ_{AGP} is well-known /36,47/. Notice that in the case M=N, with $\xi_1 = \xi_2 = \ldots\xi_i = \ldots = \xi_{N/2}$ the AGP function coincides with the single determinant $|u_1(1)\ldots u_i(i)\ldots u_N(N)$. To increase the number of determinants is the normal procedure for introducing correlation. But we have clearly another alternative: <u>to increase the rank M from N to some higher value, possibly infinite</u>. The structure of $\underline{D}^{(2)}(g^{N/2})$ is simple. It consists of a block of rank M associated to the basis $u_{2i-1}(1)u_{2i}(2)|$, no off-diagonal elements connecting this block to the rest of $\underline{D}^{(2)}$, where the matrix is diagonal. We have exploited these properties in a construction of the AGP solution and its associated propagator /33/. We used unitary transformations systematically in the way of Dalgaard and Jørgensen /2,14,57/. The orbital optimization part parallels the one in modern MC-SCF or super CI procedures /13,14,57/. The coefficient variation is not associated, of course, to the linear variation of coefficients as in MC-SCF, but rather to a simple procedure dependent on the special nature of $\underline{D}^{(2)}(g^{N/2})$.

FROM AGP TO AMO

The semilocalized orbitals of Coulson and Fischer /18/ were used by Löwdin /17/ in order to have different spins on two different subsets of atoms in an alternant system. The AMO's are obtained by coupling occupied MO's with virtual ones according to a coupling scheme /20,60, 61/. Let Ψ_K and $\Psi_{\bar{K}}$ denote such a pair - usually a bonding and an antibonding orbital. Then the AMO's are:

$$A_K = \Psi_K\cos\theta_K + \Psi_{\bar{K}}\sin\theta_K; \quad \bar{A}_K = \Psi_K\cos\theta_K - \Psi_{\bar{K}}\sin\theta_K. \tag{15}$$

The AMO's are normalized and their overlap is $\langle A_K|\bar{A}_K\rangle = \cos 2\theta_K = \lambda_K$. One constructs then a single determinant with AMO's, instead of a single determinant with doubly filled molecular orbitals Ψ_K.

$$\Psi_{AMO} = N(\lambda_1,\ldots,\lambda_{N/2})|A_1\alpha\bar{A}_1\beta,\ldots,A_K\alpha\bar{A}_K\beta,\ldots,A_{N/2}\alpha\bar{A}_{N/2}\beta|$$

This wave function should be projected in order to recover the proper spin and spatial symmetry desired, though this may not be necessary in large systems /20/. This is a wave function of the DODS type, and a special case of the unrestricted Hartree-Fock method (UHF). As shown by Kapuy, the AMO function is a special case of the separated pair theory /62/. This is also known as the antisymmetrized product of strongly orthogonal geminals (APSG), as discussed by Kutzelnigg /42/:

$$\Psi_{APSG} = A\ g_1(1,2)g_2(3,4)\ldots g_{N/2}(N-1,N) \tag{16}$$

with $\int g_i^*(1,2)g_j(2,3)dv_2 = 0$ for $i \neq j$.

The unprojected AMO function, being a single determinant can be written in the AGP form (13) and (14). One takes M = N and equal weights ξ_i:

$$g^{AMO}(1,2) = \sum_{K=1}^{N/2}|A_{2K-1}(\vec{r}_1)\alpha(\xi_1)\ \bar{A}_{2K}(\vec{r}_2)\beta(\xi_1)| \tag{17}$$

which clearly is a mixture of singlet and triplet in the geminal. The triplet component has $\langle S_z\rangle = 0$. We can express $g^{AMO}(1,2)$ in terms of the MO basis

$$g^{AMO}(1,2) = \sum_{K=1}^{N/2}\{|\Psi_K\alpha\Psi_K\beta|(\cos\theta_K)^2 + \{-|\Psi_K\alpha\Psi_{\bar{K}}\beta| + |\Psi_{\bar{K}}\alpha\Psi_K\beta|\}\frac{\sin 2\theta_K}{2}$$
$$+ |\Psi_{\bar{K}}\alpha\Psi_{\bar{K}}\beta|(\sin\theta_K)^2\} \tag{18}$$

There is possibility of correct dissociation here. Furthermore, this shows that in the MO basis the AMO procedure can be considered as enlarging the rank of the AGP geminal. This implies that in a first iteration in the optimization of the AGP energy functional one could restrict the variational freedom to 2x2 rotations (this would restrict the sum in (6) to one i and one a). Our results indicated that if desired one can restrict the AGP procedure to the following form /32-33/:

$$AGP' = C' A g_1(12)g_1(34)\ldots g_2(K,K+1)\ldots g_2(N-1,N) \quad (19)$$

where we have two (or more) strongly orthogonal geminals. The excitation operators must then be restricted to the appropriate subspaces. The APSG is an extreme case of this. In particular, for the AMO case the excitation operators should have the form (in terms of the appropriate particle-hole and hole-particle operators) (6), e.g., Bogoliubov transformations. The concepts of pairing and DODS have been discussed by Berggren and Johansson /63/. That they are interrelated was clear, but the energetic relationship that guarantees that the E_{AGP} is lower than the E_{AMO} follows from the preceding considerations. In a similar vein the E_{AGP} should be lower than the E_{UHF}. The dashed line in fig. 4-5 in the monograph by Pauncz /20/, corresponding to a one-parameter AMO calculation, is then an upper limit to the energy E_{AGP}.

The question of the many-parameter case, on how to determine the splitting parameters θ_K and their associated "gap-equation" is a highly relevant question. This is currently discussed by Calais /64/. An optimal pairing scheme of the AMO type is introduced. One can nevertheless say that this optimal pairing cannot yield as much energy lowering as the AGP function, when <u>all possible</u> pairings appear. The relevance of this work in solids and large systems has to do with charge and spin density waves, all aspects of a "localization" introducing correlation /65/. The Hubbard hamiltonian has been used by Johansson and Berggren to model Mott transitions in a one-dimensional chain /66/. The relevance of using different values of θ_K is clear from their work, and also that an AGP treatment would improve further the energy. A derivation of the empirical relation between ε_K and θ_K given by Pauncz, starting from the "gap-equation" is still missing.

For atoms and molecules the question of using the right symmetry can be essential. In the AMO method this is implemented by a projection operator /20,62/:

$$^S\Psi = {}^S O \Psi_{AMO} \quad (20)$$

In the case of singlet coupling this leads to an expansion in terms of configurations evenly excited with respect to a reference determinant. This is a feature exhibited by the AGP function in its canonical form (13) and (14). The projected AMO is not of AGP form, though its density matrix can be constructed in a simple way /20/. The work of Kapuy /62/ is useful in this connection. Spin projection of (16) will introduce a superposition of similar functions with contributions of all the possible spin couplings in the geminal. We leave this question in the present context since there are technical but no conceptual difficulties. The introduction of the $\underline{h}_4$ manifold becomes necessary. The precise nature of the relationship between AMO, the projected BCS, UHF and AGP has been hinted at before, mainly through the pairing scheme. That AMO states and the AGP state have many features in common and also a large overlap has been anticipated by Linderberg /45/.

Previous molecular applications have been made with a singlet geminal /39-42/. Both singlet and triplet coupling have been relevant in superphenomena /34-38/. It is clear that under normal circumstances there is no special reason to expect a geminal of pure spin (in an AGP for an atom or molecule).

An interesting comment is the relationship of the AGP with highly coherent phenomena including superconductivity, which, of course, has been extensively discussed in the literature. This is associated to "extreme" character of the AGP function (as emphasized by Coleman /23, 36/): all the weights in (14) equal to each other. For $M \to \infty$, there is then possibility of the largest eigenvalue of $\underline{D}^{(2)}(g^{N/2})$ to approach its upper bound, of "macroscopic" character ($\sim N$), as Sasaki, Yang and others have shown /67/. Some of our studies /68/ illustrate how the extreme AGP, with $M > N$ leads also to a gap in the excitation spectrum and a vanishing transition moment, which for certain perturbations is responsible for the "rigidity" of the BCS function in the presence of a magnetic field and hence to the Meissner effect /37/. It can be instructive to consider the following simplified model. If one has a singlet coupled geminal, for $M = 4$, (14) reduces to:

$$g = 2^{-\frac{1}{2}}[\cos\theta|1\bar{1}| + \sin\theta|2\bar{2}|] \tag{21}$$

where we have simplified notation: i and $\bar{i}$ denote spin orbitals with the same spatial part and with α and β spin respectively. The operator

$$\Omega^+ = 2^{-\frac{1}{2}}(\cos^2\theta - \sin^2\theta)^{-1}\ (\cos\theta q - \sin\theta q^+), \tag{22}$$

where $q = \overset{+}{a}_{\bar{2}}a_{\bar{1}} + \overset{+}{a}_2 a_1$ and $q^+ = \overset{+}{a}_{\bar{1}}a_{\bar{2}} + \overset{+}{a}_1 a_2$ fulfils the following relation:

$$\Omega^+ g = 2^{-1}[|1\bar{2}| + |2\bar{1}|] \qquad \equiv e;\ \Omega g = 0 \tag{23}$$

It is a proper excitation operator. For $\cos\theta = \sin\theta$ this breaks down completely. If we say that g represents H_2, $\cos\theta \sim \sin\theta$ asymptotically. The wave functions g become (factorizing the spin and neglecting overlap between atomic orbitals a and b):

$$g \sim 2^{-\frac{1}{2}}(ab+ba);\quad e \sim 2^{-\frac{1}{2}}(aa-bb) \tag{24}$$

The upper state e is ionic with a finite energy gap with respect to the ground Heitler-London state. The transition density matrix between these two vanishes for real orbitals. This is a manifestation of the consequences of $\cos\theta = \sin\theta$, i.e., of the extreme behaviour. In this example it does not have physical significance because we have imposed coherence at large distances through the pure singlet requirement. The singlet--triplet distinction ceases to be meaningful then. But, in analogous situations, through coupling to a third partner or agent, the situation can become quite realistic. The essence is the breakdown of the one--body excitation spectrum. This rigidity is an essential feature of the BCS solution and is inherent to extreme type of behaviour. This is why wave guide type of analysis of superconductivity, with emphasis on a one-particle picture near the Fermi surface may be misleading or incomplete, though pictorially very attractive /69/.

AGP AND THE THEORY OF CHEMICAL REACTIONS

It would be presumptuous to summarize or even refer here to all the work which has been done on the theory of chemical reactions and on the associated electronic surfaces /70/. The emphasis placed by Fukui on amplitude and nodal pattern of particular changes /70/ is related to Halevi's orbital correspondence /71/ and to the omnipresent orbital or state correlation diagrams of Woodward and Hoffman (WH) and of Longuet-Higgins and Abrahamson /72/. An essential aspect of the well-known correlation diagrams for e.g. orbital energies, is that occupied HF orbitals correlate with virtual ones along the reaction path in a thermally "forbidden" reaction. In a very significant paper, Wilson and Wang made the observation that the correlation diagram in terms of occupation numbers of natural orbitals for a WH "forbidden" reaction should have a rather different character from the one associated to an "allowed" reaction /73/. Their analysis regarded, in principle, the exact wave function, where under normal circumstances the occupation numbers of the natural orbitals are close to zero or to one /74/. In an "allowed" reaction the occupation numbers do not change order of magnitude. This exact treatment as such is not very useful. One needs an approximate model which can accommodate the possibility of changes of configuration, of changes of occupation numbers. A step in this direction is the recent work by Ruedenberg /75/. They provide an orbital reaction space and a configurational reaction space leading to "natural reaction orbitals" (NRO). The first order reduced spinless density matrix has the form

$$\rho = \sum_{j} 2\phi_i\phi_i^* + \sum_{K} n_K \Psi_K \Psi_K \tag{25}$$

The essential point is that the n_K's and the Ψ_K's change during the reaction within the multiconfigurational HF prescriptions. This very pictorial description can be considered within the AGP formalism. One would include the active orbitals in the construction of g, as in (13) and (14) and then the construction of the first order density matrix in forms like (25) would follow. AMO approximations would be simple pairs of orbitals interacting à la Coulson-Fischer. More sophisticated schemes, like the UHF ones of Fukutome /76/. An essential result in this context is that the natural spin orbitals of g and of Ψ_{AGP} coincide /36/. The AGP wave functions are in a way the simplest possible extension of HF theory. In the latter the first order reduced density matrix determines all the higher ones, in particular the second order one. In the former the natural orbitals and their occupation numbers determine g, except the phase factors ξ_i in (14). Notice that the first order density matrix is of

$$\Gamma^{(1)}(1,1') = (N/2)\sum_{i=1}^{M/2} |\xi_i|^2 \{u_{2i-1}(1)u_{2i-1}^*(1') + u_{2i}(1)u_{2i}^*(1')\}. \tag{26}$$

One can monitor then changes in the second order density matrix, "caused" by changes in the first order one, through the changes in g.

CORRELATION EFFECTS THROUGH BROKEN SYMMETRY SOLUTIONS

It has been shown by many authors that correlation corrections to the HF solution may be introduced by lowering the symmetry constraints, facing the symmetry dilemma in HF theory /77/. The pioneering work was carried out by Bagus and Schaeffer /78/. Recent contributions involve core and valence ionization and n-π^* transitions /79-83/, not mentioning at all extended systems. There is no question that for core-holes the localized, broken symmetry, solutions represent quite adequately the situation. Symmetry projection is not expected to introduce an appreciable splitting - at any rate larger than the experimental accuracy /84/. In low-lying excitations this is not entirely impossible and in valence ionization it is expected to be the case, as our experience with the broken symmetry solutions of pyrazine indicate /82,85/. The use of broken orbital symmetry was essential at the computational level used. This is a vast subject and the time and space available does not allow a thorough discussion here. We can consider two examples which illustrate the power and the failures of broken symmetry solutions. Changes of bond length in homonuclear diatomics (leading to vibrational broadening of ESCA core lines /86/) indicate that not only the relaxation energy is significant, but also its derivative (relaxation energy being the correction to Koopmans's approximation due to orbital readjustment) /87/.

As discussed elsewhere /84/ the need to project depends on the nature of an effective coupling constant. In a diatomic this depends on the distance. Calculations on N_2 by Bartlett and Purvis illustrate this point /88/. UHF calculations on the ground state of N_2 exhibit a "bump" (due to the contribution of higher energy triplet, quintet and septet states) for an appreciable region to the right of the onset of the Coulson-Fischer instability. This anomalous behaviour persists even in the correlated wave function obtained from the UHF solution. These questions, and some of the developments on the use of many-body perturbation theory (MBPT) were discussed extensively in a recent symposium /10/.

EVALUATION AND PERSPECTIVES

In a piercing analysis of the quantum chemical world in 1970, Mme. Pullman with both esprit and insight took up the conflict between those for which quantum chemistry is essentially chemistry, studied and deepened by quantum mechanical methods and those for which it is essentially a framework, if not an excuse, for the development and the exploration of computational methods /89/. One may add to this, perhaps in defence of the second cathegory, that the development of formal and computational methods provides links with other parts of science (and within our subject itself preventing that "dialects" become too isolated). It is mutually beneficial for nuclear-, solid- and surface-physics to interact with chemistry. Formal methods provide the language of communication. An example of how useful it is to have a continuous hunt for methods and new possibilities is the independent development of group theory, and in particular of unitary groups /15/, which provides the framework for modern computational methods. An essential aspect in this connection is the recognition that the hamiltonian in second quantization can be expressed in terms of the generators of a commutator algebra /15/. Algebras of this type have been extensively used by Matsen /90/. In propagator theory they acquire a natural role through equations of the form

$$\hat{H}Q_K^+ = [Q_K^+, H] = (E_K - E_o)Q_K^+$$

We may expect that even the theory for the general linear group, by Moshinsky and Seligman /91/, will provide a similar development for non-orthogonal orbitals. Preliminary applications to AMO have been made already /92/. In a similar vein, the "geometry" of density matrices /93/ will be exceedingly useful.

A long time has elapsed since Shull coined the word geminal and group functions were analyzed by McWeeny /94/. We anticipate today a variety of geminal functions which may condense somewhat the multitudinous character that CI expansions have. Of interest in this context is to construct the appropriate ground state for excitation operators involving $\underline{h}_2$ and $\underline{h}_4$ (necessary for spin-adaption as well as correlation) as well as those involving $\underline{h}_1$ and $\underline{h}_3$ (necessary for shake-up and shake--off processes). The methods proposed here do not exclude other ones, e.g., involving MBPT as carried out by Cederbaum et al. /5,6/. They yield though possibility of analysis and interpretation in terms of orbital concepts and models.

We expect that temperature effects, environmental ones, vibronic coupling, internal energy transfer in large molecules can be handled efficiently by propagator methods. An example of this is the work by Ratner, using the particle-hole propagator, on intramolecular electron transfer processes /95/.

Good account of photoelectron spectroscopy can be made with current techniques. The MO ionization picture found to be valid for both outer valence and core electrons may break down in the inner valence region /96/. Recent advances using superoperators /97/ indicate the power of the method for generating the detailed cancellations involved in the calculations of excitation energies. In a similar way, the use of transition orbitals in perturbative expressions for ionization energies /98/ indicate that the transition orbitals yield high accuracy in both relaxation and correlation corrections at low order.

The distinction between semiempirical vs *ab initio* becomes less controversial in second quantization. Propagators can be modelled with "less violence" than wave function methods.

The use of the electron propagator in conjunction with the fastly developing area of photoelectron spectroscopy will certainly continue, as well as the polarization propagator in connection with excitation processes. A recent example is the work by Karplus which is directly related to the gist of this paper. It illustrates some of the theoretical difficulties, and to some extent the consequences of using perturbation theory, and the need of solutions beyond CI character, perhaps of the type suggested here.

It might be appropriate to emphasize that there is similitude with the MC-SCF procedure (as carried out by Siegbahn et al. /14/) insofar the orbital optimization. Their coefficient optimization is linear. The coefficients in the eventual expansion of an AGP function in terms of determinants are non-linearly related and hence $E_{MC-SCF} \leqslant E_{AGP}$ when one uses the same one-electron basis. The price of having a worse energy is hopefully paid by having an excitation spectrum generated in an optimal way. The excited states in MC-SCF theory would be higher roots of the secular equation with no formal connection with the ground state except through N-electron excitation operators which are approximated in a non-systematical way. The connection would be through numerical results. It would seem that one needs flexibility in the treatment. The world of "l'ab initio pour tous" is not monolithic.

The e^- - $2e^-$ experiments currently done start unravelling the details of the second order density matrix. The two-electron propagator will be the appropriate tool for the direct theoretical description of the e - $2e^-$ experiments /99/.

GLOSSARY

One-Electron Propagator

$$\langle\langle a_K^+, a_e \rangle\rangle_E$$

(poles at ionization energies and electron affinities)

Particle-Hole Propagator

$$\langle\langle a_a^+ a_i, a_b^+ a_j \rangle\rangle_E$$

(poles at excitation energies)

Two-Electron Propagator

$$\langle\langle a_K^+ a_{K'}^+ ; a_e a_{e'} \rangle\rangle_E$$

(poles at double ionization energies and two-electron affinities)

They are related through contour integration (with contributions from the various residues) to the following elements of reduced density matrices /2/:

$$\underline{D}^{(1)}_{eK} = \langle 0|a^{+}_{K}a_{e}|0\rangle$$

$$\underline{D}^{(2)}_{e'eKK'} = \langle 0|a^{+}_{K}a^{+}_{K'}a_{e}a_{e'}|0\rangle.$$

ACKNOWLEDGEMENTS

I wish to thank the organizers for their generous invitation, the Swedish Natural Science Research Council for support and Dr. Brian Weiner for fruitful and pleasant collaboration.

REFERENCES

1. Y. Öhrn, in *The New World of Quantum Chemistry*, Proceedings from the 2nd International Congress of Quantum Chemistry (B. Pullman and R. Parr, Eds., D. Reidel, Boston, 1976) pp. 57.

2. J. Linderberg and Y. Öhrn, *Propagators in Quantum Chemistry* (Academic Press, London, 1973)

3. P. Jørgensen, Ann. Review of Physical Chem. 26, 359 (1975).

4. J. Oddershede, Advances in Quantum Chem. 11, 275 (1978).

5. L.S. Cederbaum and W. Domcke, Advances in Chemical Phys. 36, 205 (1977).

6. W. von Niessen, L.S. Cederbaum and W. Domcke in *Excited States in Quantum Chemistry* (C. Nicolaides and D.R. Beck, Eds., D. Reidel, Dordrecht, 1979) pp. 183.

7. Y. Öhrn in *Excited States in Quantum Chemistry* (C. Nicolaides and D.R. Beck, Eds., D. Reidel, Dordrecht, 1979) pp. 317.

8. C.W. McCurdy, T.N. Rescigno, D.L. Yeager and W. McKoy in *Modern Theoretical Chemistry* (H.F. Schaefer III, Ed., Plenum Press, New York, 1977) pp. 349.

9. M.F. Herman, D.L. Yeager and K.F. Freed, Chem. Phys. 29, 77 (1978); D.L. Yeager and K.F. Freed, Chem. Phys. 22, 415 (1977).

10. For a recent discussion see proceedings of the Nobel Symposium 46, *Many Body of Atomic Systems*, Edited by I. Lindgren and S. Lundqvist, Physica Scripta (to appear).

11. C.A. Coulson in *Proceedings of Int. Conference of Theoretical Physics* (Kyoto and Tokyo 1953, Science Council of Japan Publ., Ueno

Park, Tokyo, 1954).

12. B. Roos and P.E.M. Siegbahn in *Modern Theoretical Chemistry 3: Methods of Electronic Structure Theory* (H.F. Schaefer III, Ed., Plenum Press, New York, 1977)

13. F. Grein and T.C. Chang, Chem. Phys. Lett. 12, 44 (1971); A. Bannerjee and F. Grein, Int. J. Quantum Chem. 10, 123 (1976).

14. P. Siegbahn, A. Heiberg, B. Roos and B. Levy, Physica Scripta 00, 000 (1980).

15. This is a vast subject, going back to Weyl, Gelfand-Zhitlin and Moshinsky. For references and recent work see J. Paldus in *Theoretical Chemistry: Advances and Perspectives* (H. Eyring and D.J. Henderson, Eds., Academic Press, New York, 1976) Vol. 2, pp. 31; I. Shavitt, Int. J. Quantum Chem. Symp. 11, 131 (1977); ibid, Symp. 12, 5 (1978).

16. J.C. Slater, Phys. Rev. 35, 509 (1930); *Proc. Shelter Island Conf. on Quantum Mechanical Methods in Valence Theory* (1951) pp. 121.

17. P.O. Löwdin, *Proceedings of the Nikko Symposium on Molecular Physics* (M. Kotani et al., Eds., Maruzen Co., Tokyo, 1954) pp. 13.

18. C.A. Coulson and I. Fischer(-Hjalmars), Phil. Mag. 40, 386 (1949).

19. T. Itoh and H. Yoshizumi, J. Phys. Soc. Japan 10, 201 (1955); J. Chem. Phys. 23, 412 (1955).

20. Many aspects of the AMO method are aptly discussed by R. Pauncz in *Alternant Molecular Orbital Method* (W.B. Saunders Co., Philadelphia, 1967), where an extensive bibliography is also given.

21. See e.g. D.J. Thouless, *The Quantum Mechanics of Many Body Systems* (Academic Press, New York, 1961).

22. P.O. Löwdin, J. Phys. Chem. 61, 55 (1957).

23. A.J. Coleman, Rev. Mod. Phys. 35, 668 (1963).

24. J. Linderberg and Y. Öhrn, Int. J. Quantum Chem. 12, 161 (1977).

25. Y. Öhrn and J. Linderberg, Int. J. Quantum Chem. 15, 343 (1979).

26. H. Fukutome, M. Yamamura and S. Nishiyama, Progr. Theoret. Phys. 57, 1554 (1977); H. Fukutome, Progr. Theoret. Phys. 58, 1692 (1977).

27. O. Goscinski and B. Lukman, Chem. Phys. Letters 7, 573 (1970).

28. B.T. Pickup and O. Goscinski, Mol. Phys. 26, 1013 (1973).

29. B.T. Pickup and O. Goscinski, Chem. Phys. Letters 33, 265 (1975).

30. B. Weiner and O. Goscinski, Int. J. Quantum Chem. 12, 299 (1977).

31. O. Goscinski and B. Weiner, Physica Scripta 00, 000 (1980).

32. B. Weiner and O. Goscinski, *The Self-Consistent Approximation to the Polarisation Propagator*, submitted to the Int. J. Quantum Chem.

33. B. Weiner and O. Goscinski, *Calculation of Optimal Antisymmetrized*

Power (projected-BCS) Functions and their Associated Excitation Spectrum, to be submitted to the Phys. Rev.

34. L. Bardeen, L.M. Cooper and J.R. Schrieffer, Phys. Rev. 108, 1175 (1957).
35. J.M. Blatt, *Theory of Superconductivity* (Academic Press Inc., New York, 1964).
36. A.J. Coleman, J. Math. Phys. 6, 1425 (1965).
37. J.R. Schrieffer, *Theory of Superconductivity* (W.A. Benjamin, Ed., New York, 1964).
38. J. Leggett, Rev. Mod. Phys. 47, 331 (1975).
39. S. Bratoz and Ph. Durand, J. Chem. Phys. 43, 2670 (1965).
40. G. Bessis, C. Murez and S. Bratoz, Int. J. Quantum Chem. 1, 327 (1967).
41. G. Bessis, P. Espagnet and S. Bratoz, Int. J. Quantum Chem. 3, 205 (1969).
42. W. Kutzelnigg, J. Chem. Phys. 40, 3640 (1964).
43. R.E. Watson, Phys. Rev. 119, 170 (1960).
44. W. Kutzelnigg, Theoret. Chim. Acta (Berl.) 3, 241 (1965).
45. J. Linderberg, Israel J. of Chemistry 00, 000 (1980).
46. P.O. Löwdin, Phys. Rev. 139, 357 (1965).
47. M. Rosina, in Queen's Papers in Pure and Applied Math. No. 40, Kingston, Ontario, Canada (1974).
48. J. Linderberg, Physica Scripta 00, 000 (1980).
49. T.A. Koopmans, Physica 1, 104 (1933).
50. R.S. Mulliken, J. Chem. Phys. 46, 497 (1949).
51. P.S. Bagus, Phys. Rev. 139, 619 (1965).
52. As an example, rather than an extensive bibliography see: *Excited States in Quantum Chemistry* (C.A. Nicolaides and D.R. Beck, Eds., D. Reidel, Dordrecht, 1979).
53. A.D. McLachlan and M.A. Ball, Rev. Mod. Phys. 36, 844 (1964) and references therein.
54. J. Linderberg, Arkiv f. Fysik 26, 232 (1964).
55. P. Lindner and O. Goscinski, Int. J. Quantum Chem. 4S, 251 (1951).
56. A. Calles and O. Goscinski, *The Use of Green Functions for the Calculations of the Dynamic Jahn-Teller Effect*, preprint.
57. P.O. Löwdin, Int. J. Quantum Chem. 2, 867 (1968).
58. E. Dalgaard, Int. J. Quantum Chem. 15, 197 (1979); E. Dalgaard and P. Jørgensen, J. Chem. Phys. 69, 3833 (1978); R. Manne, Chem. Phys.

Letters 45, 470 (1977).

59. C. Nehrkorn, G. Purvis, Y. Öhrn, J. Chem. Phys. 64, 1752 (1978).

60. R. Pauncz, J. de Heer and P.O. Löwdin, J. Chem. Phys. 36, 2247, 2257 (1962).

61. O. Goscinski and J-L Calais, Arkiv Fysik 29, 135 (1965).

62. E. Kapuy, Theoret. Chim. Acta 3, 379 (1965). See also I. Mayer, Acta Physica Acad. Scient. Hungarical, T37, 39 (1974); ibid, T34, 305 (1973).

63. K.F. Berggren and B. Johansson, Int. J. Quantum Chem. 2, 483 (1968). See also Physica 40, 277 (1968).

64. J-L Calais, Ann. Soc. Scient. Bruxelles T93, 000 (1979). See also *Mott and Peierls Gaps*, TN 603, Uppsala Quantum Chemistry Group, 1979, and references therein.

65. A.A. Ovchinnikov, J.J. Ukrainskii and G.V. Kventsel, Soviet Phys. Uspekhi 15, 575 (1973).

66. B. Johansson and K.F. Berggren, Phys. Rev. 181, 855 (1969).

67. F. Sasaki, Phys. Rev. 138, B1338 (1965); C.N. Yang, Rev. Mod. Phys. 35, 668 (1963); A.J. Coleman, Int. J. Quantum Chem. 13, 67 (1978).

68. B. Weiner and O. Goscinski, to be published.

69. K.H. Johnson, D.D. Vvedensky and R.P. Messmer, *Molecular Orbitals and Superconductivity* (preprint).

70. Consider the review by K. Fukui, Int. J. Quantum Chem. 12, Suppl. 1, 277 (1977) and references therein as an example.

71. E.A. Halevi, Int. J. Quantum Chem. 12, Suppl. 1, 289 (1977).

72. R.B. Woodward and R. Hoffmann, J. Am. Chem. Soc. 87, 395 (1965); H.C. Longuet-Higgins and E.W. Abrahamson, J. Am. Chem. Soc. 87, 2045 (1965).

73. E.B. Wilson and P.S.C. Wang, Chem. Phys. Letters 15, 400 (1972).

74. E.R. Davidson, *Reduced Density Matrices in Quantum Chemistry* (Academic Press, New York, 1976).

75. K. Ruedenberg and K.R. Sundberg in *Quantum Science* (J-L Calais et al., Eds., Plenum Press, New York, 1976) pp. 505.

76. An interesting series of papers relating chemical reactivity with HF instabilities and broken symmetry solutions: H. Fukutome and coworkers, Prog. Theoret. Phys. 47, 1156 (1972); 49, 22 (1973); 50, 1433 (1973); 52, 1580 (1975); 54, 1599 (1975).

77. P.O. Löwdin, Rev. Mod. Phys. 35, 496 (1963).

78. P.S. Bagus and H.F. Schaeffer III, J. Chem. Phys. 56, 224 (1972).

79. H.T. Jonkman, G. Van der Velde and W.C. Nieuwpoort, SRC Atlas Symposium No. 4, Oxford (1974).

80. L.E. Nitzsche and E.R. Davidson, Chem. Phys. Letters 58, 171 (1978).

81. C.P. Keijzers, P.S. Bagus and J.P. Worth, J. Chem. Phys. 69, 4032 (1978).

82. S. Canuto, O. Goscinski and M. Zerner, Chem. Phys. Letters 00, 000 (1979).

83. J. Müller, E. Poulain and L. Karlsson, J. Chem. Phys. 00, 000 (1980).

84. R.L. Lozes, O. Goscinski and U.I. Wahlgren, Chem. Phys. Letters 63, 77 (1979).

85. S. Canuto and O. Goscinski, Uppsala Quantum Chemistry Group, TN 599 (1979).

86. J. Müller, H. Ågren and O. Goscinski, Chem. Phys. 38, 349 (1979) and references therein to work by Clark and collaborators.

87. O. Goscinski and A. Palma, Chem. Phys. Letters 47, 322 (1977); A. Palma et al., Chem. Phys. Letters 62, 368 (1979).

88. R.J. Bartlett and G.D. Purvis, Physica Scripta 00, 000 (1980).

89. A. Pullman in *Aspects de la Chimie Quantique Contémporaine*, Colloques Int. du CNRS 195 (Editions du CNRS, Paris 1970) pp. 9.

90. F.A. Matsen, Int. J. Quantum Chem. 10, 525 (1976); Adv. Quantum Chem. 11, 223 (1978).

91. M. Moshinsky and T.H. Seligman, Ann. Phys. 66, 311 (1971).

92. C.R. Sarma and K.V. Dinesha, Int. J. Quantum Chem. 15, 579 (1979).

93. J.E. Harriman, Int. J. Quantum Chem. 15, 611 (1979).

94. H. Shull, J. Chem. Phys. 30, 1405 (1957); R. McWeeny and Y. Mizuno, Proc. Roy. Soc. (London) A259, 554 (1961).

95. M.A. Ratner, Int. J. Quantum Chem. 5, 675 (1978).

96. L.S. Cederbaum, J. Schirmer, W. Domcke and W. von Niessen, Int. J. Quantum Chem. 14, 593 (1978).

97. G. Born and Y. Öhrn, Chem. Phys. Letters 61, 307 (1979).

98. H.A. Kurtz and Y. Öhrn, J. Chem. Phys. 69, 1162 (1978).

99. V.G. Neudatchin et al., Phys. Lett. 64A, 31 (1977).

THE UNITARY GROUP FORMULATION OF THE MANY-ELECTRON PROBLEM†

F. A. Matsen and C. J. Nelin
Departments of Physics and Chemistry
The University of Texas
Austin, Texas 78712

ABSTRACT

The unitary group formulation is a viable alternative to the Slater determinant and second quantized methods when the Hamiltonian is spin-free. Here the Hamiltonian is expressed as a second degree polynomial in the infinitesimal generators of $U(\rho)$ where ρ is the number of spatial (spin-free) orbitals. Each irreducible space of $U(\rho)$ is invariant under this Hamiltonian and is uniquely characterized by a Young diagram which, for the Pauli allowed spaces, supplies both particle and spin quantum numbers. Each irreducible space is spanned by a set of orthonormal Gel'fand states and the Hamiltonian matrix elements are conveniently calculated by the graphical unitary approach (GUGA). The many-body approximation theories (single and multiconfiguration restricted Hartree Fock, RPA, perturbation, coupled cluster and effective Hamiltonian theories) have been given a unitary group formulation

I. INTRODUCTION

The Slater determinant and the second-quantized formulations are standard procedures in molecular quantum mechanics. In 1974 a third procedure, the unitary group formulation, was introduced [1] based on earlier work of Gel'fand [2], Moshinsky [3] and Biedenharn [4]. Its advantages are as follows: (i) It does not require spin projection. (ii) Matrix elements are easy to evaluate [5,6]. (iii) It provides a good basis for many-body theory [7]. (iv) For a spin-free Hamiltonian the theory is spin-free [8,9].

II. A MODEL HAMILTONIAN [1]

Our model Hamiltonian is defined in terms of a set of ρ orthonormal, spin-free orbitals

$$V(\rho): \quad \{|r\rangle, \; r=1 \text{ to } \rho\} \tag{2.1}$$

†Supported by R. A. Welch Foundation of Houston, Texas

K. Fukui and B. Pullman (eds.), Horizons of Quantum Chemistry, 37–48.

For an N-electron system (before the application of the Pauli exclusion principle) the vector space is the product space

$$V(\rho^N): \quad \{|r(N)> \equiv |r_1>|r_2>\cdots|r_N>, \ \dim = \rho^N\} \tag{2.2}$$

where the subscript is an electron index.

The $\rho^N \times \rho^N$ representation of the Schroedinger Hamiltonian on $V(\rho^N)$ is denoted by

$$[H] = [<r(N)|H|r'(N)>] \tag{2.3}$$

and is exactly modeled by

$$H = \sum_{mrs}\sum\sum |r_m>h_{rs}<s_m| + \tfrac{1}{2}\sum_{m\neq n}\sum_{rstu}\sum\sum\sum |t_n>|r_m>v_{rs,tu}<s_m|<u_n| \tag{2.4}$$

where h_{rs} $(\equiv<r_m|h_m|s_m>)$ and $v_{rs,tu}$ $(\equiv<t_n|<r_m|g_{mn}|s_m>|u_n>)$ are parameters which are independent of the particle index. They can be evaluated ab initio by integration over a carefully selected set of analytical orbitals or chosen semiempirically. On substitution we have

$$H = \sum_{rs}\sum h_{rs}\sum_m |r_m><s_m| + \tfrac{1}{2}\sum_{rstu}\sum\sum\sum v_{rs,tu}\{\sum_{m\neq n}(|r_m><s_m|)(|t_n><u_n|)\} \tag{2.5}$$

The curly bracket is rewritten as follows:

$$\begin{aligned}
&\sum_{m\neq n}(|r_m><s_m|)(|t_n><u_n|)\\
&= (\sum_m|r_m><s_m|)(\sum_n|t_n><u_n|) - \sum_m(|r_m><s_m|)(|t_m><u_m|)\\
&= (\sum_m|r_m><s_m|)(\sum_n|t_n><u_n|) - \sum_m(\delta_{s_m t_m}|r_m><u_m|)
\end{aligned} \tag{2.6}$$

Now define

$$E_{rs} \equiv \sum_m |r_m><s_m| \tag{2.7}$$

and we have for the model Hamiltonian on $V(\rho^N)$,

$$H = \overline{\sum_{rs}\sum}^{\rho} h_{rs}E_{rs} + \tfrac{1}{2}\overline{\sum_{rstu}\sum\sum\sum}^{\rho} v_{rs,tu}(E_{rs}E_{tu} - \delta_{st}E_{ru}) \tag{2.8}$$

Note that as a consequence of the orthonormality of the orbitals

$$[E_{rs},E_{tu}] = \delta_{st}E_{ru} - \delta_{ru}E_{ts} \tag{2.9}$$

III. UNITARY GROUP THEORY [10,11]

The group

$$U(\rho) = \{x, y, z, \cdots\}, \quad x^{\dagger} = x^{-1} \tag{3.1}$$

is the set of all unitary transformations of a ρ dimensional carrier space (2.1). The group elements have exponential forms,

$$x = e^{X} \text{ where } X^{\dagger} = -X \text{ (skew Hermitian)} \tag{3.2}$$

The Lie algebra is

$$LAU(\rho) = \{X, Y, Z, \cdots\} \tag{3.3}$$

which is closed under the Lie product (commutator) and is non-associative. It has a basis

$$LAU(\rho): \quad \{E_{rs}; \; r,s = 1 \text{ to } \rho\}, \; E^{\dagger}_{rs} = E_{sr} \tag{3.4}$$

with

$$X = \sum_{rs}^{\rho}\sum X_{rs} E_{rs}, X^{*}_{rs} = -X_{rs} \in ¢ \tag{3.5}$$

The Lie products are defined by (2.9). Note that

$$\lim \frac{\partial x}{\partial X_{rs}} \xrightarrow[X\to 0]{} E_{rs} \text{ (an infinitesimal generator of } U(\rho)) \tag{3.6}$$

The covering algebra is

$$CAU(\rho): \quad \{I, E_{rs}, E_{rs}\, E_{tu}, \cdots\} \equiv E(i,\tau) \tag{3.7}$$

where i is the degree so $x \in CAU(\rho)$. The group space is decomposed according to

$$V(U(\rho)) = \sum_{[\lambda]} \oplus V_{\rho}[\lambda] \tag{3.8}$$

where each $V_{\rho}^{[\lambda]}$ is invariant under E_{rs} and where

$$[\lambda] = [\lambda_1, \lambda_2, \cdots \lambda_{\rho}], \; \sum_i^{\rho} \lambda_i = k = 0, 1, 2, 3 \cdots \tag{3.9}$$

is a partition of the integer k and is graphically realized by a Young diagram YD[λ] an array of k boxes with λ_1 boxes in the first row, etc. See Table 1. Each invariant space is spanned by $f_{\rho}[\lambda]$ Gel'fand states labelled by Gel'fand tableaux constructed by adding integers to YD[λ] in nondescending order along rows and descending order down columns and there exist explicit formulae for the evaluation of matrix elements of generators between Gel'fand states [2, 3, 4].

IV. THE UNITARY GROUP FORMULATION OF MANY-ELECTRON THEORY

The unitary group formulation of many-electron theory consists of combining Sections 3 and 4 and then applying the Pauli exclusion principle:

i) H (the generator Hamiltonian) (2.8) ε CAU(ρ).

ii) As a consequence of (i), $V_\rho[\lambda]$ is invariant under H and $[\lambda]$ becomes a quantum number.

iii) $[\lambda]$ is a partition of a positive integer which we set equal to the electron number, N.

iv) The Pauli exclusion principle limits YD$[\lambda]$ to no more than two columns.

v) The spin of a Pauli allowed space is equal to one-half of the difference in the lengths of the two columns. See Table 1.

vi) The Gel'fand tableaux for Pauli allowed spaces

$$V_\rho^{[\lambda]}: \ \{|F\rangle, \ F = 1 \text{ to } f_\rho^{[\lambda]}\} \tag{4.1}$$

have no more than two columns so that the orbital occupation number, $g_r \leq 2$. See Table 1.

vii) Computation of matrix elements is facilitated by the use of Paldus array [5] and the case array [8](See Table 1), which are represented in a compact form by Shavitt's distinct row table [8] and the Shavitt graph where each Gel'fand state is represented by a distinct walk (See Table 2).

viii) The matrix elements $\langle F|H|F'\rangle$ are given in closed form as a function of the walks of $|F\rangle$ and $|F'\rangle$ on a Shavitt graph. This formulation which has developed into highly efficient computing scheme is called GUGA (graphical unitary group approach) [12].

ix) viii yields an $f_\rho^{[\lambda]} \times f_\rho^{[\lambda]}$ representing $[H]^{[\lambda]}$ which is then diagonalized by conventional means.

x) $[H]^{[\lambda]}$ is exactly modeled by a <u>Gel'fand state Hamiltonian</u>

$$H = \sum_{FF'}\sum (H^0_{FF'} + V_{FF'})|F\rangle\langle F'| \tag{4.2}$$

Here

$$H^0_{FF'} = \sum_{ij}\sum h_{ij} a^{ij}_{FF'} \tag{4.3}$$

and

$$V_{FF'} = \sum_{ijkl}\sum\sum\sum v_{ij,kl} b^{ijkl}_{FF'} \tag{4.4}$$

where $a^{ij}_{FF'}(\equiv\langle F|E_{ij}|F'\rangle)$ and $b^{ijkl}_{FF'}(\equiv\tfrac{1}{2}\langle F|E_{ij}E_{kl} - \delta_{kj}E_{il}|F'\rangle)$ are called unitary group (or spin coupling) coefficients [12].

Table I. An Example $N = \rho = 3$

$[\lambda]$	$YD[\lambda]$	Spin
[3]	$\square\square\square$	excluded
[2,1]	$\begin{array}{ll}\square&\square\\ \square&\end{array}$	1/2
$[1^3]$	$\begin{array}{l}\square\\ \square\\ \square\end{array}$	3/2

Gel'fand Tableaux

$$V_3^{[2,1]} : \left\{ \begin{array}{|c|c|}\hline 1&1\\ \hline 2\\ \cline{1-1}\end{array}, \begin{array}{|c|c|}\hline 1&1\\ \hline 3\\ \cline{1-1}\end{array}, \begin{array}{|c|c|}\hline 1&2\\ \hline 2\\ \cline{1-1}\end{array}, \begin{array}{|c|c|}\hline 1&2\\ \hline 3\\ \cline{1-1}\end{array}, \begin{array}{|c|c|}\hline 1&3\\ \hline 2\\ \cline{1-1}\end{array}, \begin{array}{|c|c|}\hline 1&3\\ \hline 3\\ \cline{1-1}\end{array}, \begin{array}{|c|c|}\hline 2&2\\ \hline 3\\ \cline{1-1}\end{array}, \begin{array}{|c|c|}\hline 2&3\\ \hline 3\\ \cline{1-1}\end{array} \right\}$$

Paldus Arrays

$$V_3^{[2,1]} : \left\{ \begin{bmatrix}1&1&1\\1&1&0\\1&0&0\\0&0&0\end{bmatrix}, \begin{bmatrix}1&1&1\\1&0&1\\1&0&0\\0&0&0\end{bmatrix}, \begin{bmatrix}1&1&1\\1&1&0\\0&1&0\\0&0&0\end{bmatrix}, \begin{bmatrix}1&1&1\\1&0&1\\0&1&0\\0&0&0\end{bmatrix}, \begin{bmatrix}1&1&1\\0&2&0\\0&1&0\\0&0&0\end{bmatrix}, \begin{bmatrix}1&1&1\\0&1&1\\0&1&0\\0&0&0\end{bmatrix}, \begin{bmatrix}1&1&1\\1&0&1\\0&0&1\\0&0&0\end{bmatrix}, \begin{bmatrix}1&1&1\\0&1&1\\0&0&1\\0&0&0\end{bmatrix} \right\}$$

Case Arrays

$$V_3^{[2,1]} : \left\{ \begin{bmatrix}0\\1\\3\end{bmatrix}, \begin{bmatrix}1\\0\\3\end{bmatrix}, \begin{bmatrix}0\\3\\1\end{bmatrix}, \begin{bmatrix}1\\2\\1\end{bmatrix}, \begin{bmatrix}2\\1\\1\end{bmatrix}, \begin{bmatrix}3\\0\\1\end{bmatrix}, \begin{bmatrix}1\\3\\0\end{bmatrix}, \begin{bmatrix}3\\1\\0\end{bmatrix} \right\}$$

Table 2. $N = \rho = 3$

Distinct Row Table

i	j	a	b	c	k_0	k_1	k_2	k_3	y_0	y_1	y_2	y_3	x
3	1	1	1	1	1	2	3	4	0	2	5	6	8
2	1	1	1	0		1		2		0		1	2
	2	1	0	1	1		2	3	0		1	2	3
	3	0	2	0		2				0			1
	4	0	1	1	2	3			0	1			2
1	1	1	0	0				1				0	1
	2	0	1	0		1				0			1
	3	0	0	1	1				0				1
0	1	0	0	0									1

Shavitt Graph

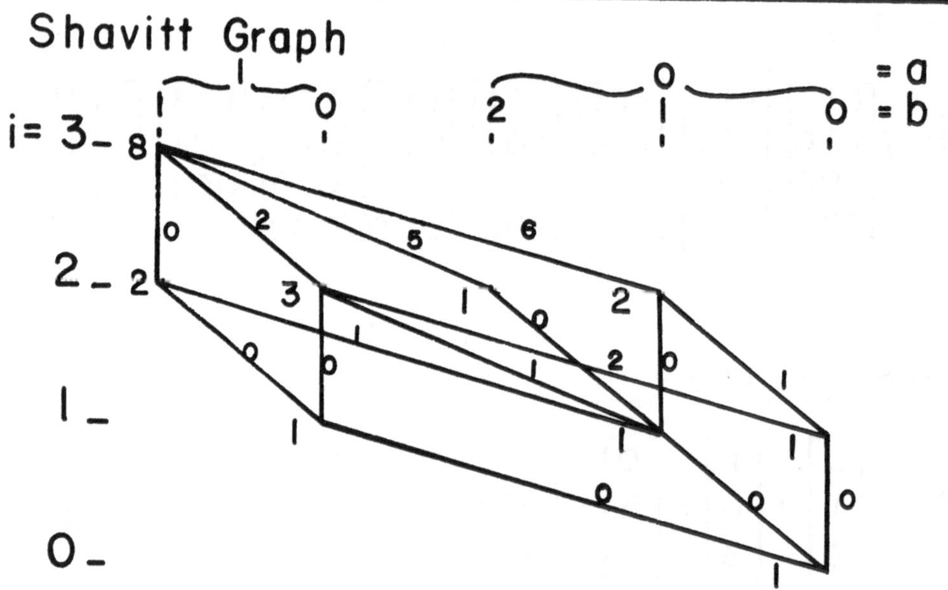

V. APPLIED MANY-ELECTRON THEORY

The unitary group formulation of many-electron theory supplies an efficient calculus and automatic spin projection:

A. Hartree Fock Theory [7,13]

We denote by $|>$ a closed shell HF ground Gel'fand state and by h (holes) and p (particles) those orbitals which are occupied and unoccupied respectively. For the variation function [13,14] we take

$$|v> = e^T|>, \quad T \equiv \sum_{ph}\sum T_{ph}(E_{ph} - E_{hp}) = -T^\dagger \tag{5.1}$$

The energy of $|v>$ is, by the BCH expansion

$$\begin{aligned} E(v) = <v|H|v> &= <|e^{-T}He^{T}|> \\ &= <|H|> + <|[H,T]|> + \tfrac{1}{2} <|[[H,T],T]|> + \cdots \end{aligned} \tag{5.2}$$

Here we employ the generator Hamiltonian and evaluate matrix elements algebraically. The second term yields the first-order extremum condition

$$<|[H,E_{ph}]|> = 0 \qquad \text{(Brillouin's theorem)} \tag{5.3}$$

and the Fock operator

$$F = \sum_{ij}^{p+h}\sum(h_{ij} + v_{ij})E_{ij}, \quad v_{ij} = \sum_h(2v_{ij,hh} - v_{ih,hj}) \tag{5.4}$$

The HF orbitals are obtained by diagonalizing F to self-consistency on the orbital space. The third term in (5.2) yields the HF stability condition, which is that $\omega \geq 0$ in

$$\begin{pmatrix} A & B \\ B^* & A^* \end{pmatrix}\begin{pmatrix} X \\ Y \end{pmatrix} = \omega\begin{pmatrix} K_{xx} & 0 \\ 0 & K^*_{yy} \end{pmatrix}\begin{pmatrix} X \\ Y \end{pmatrix} \tag{5.5}$$

where $K_{xx} = K^*_{yy} = 1$ and

$$\begin{aligned} A_{ph,p'h'} &\equiv <|[[E_{hp},H],E_{p'h'}]|> \\ &= \delta_{hh'}\delta_{pp'}2(\varepsilon(p)-\varepsilon(h)) + 2(2v_{h'p',ph} - v_{h'h,pp'}) \\ B_{ph,p'h'} &\equiv -<|[[E_{hp},H], E_{h'p'}]|> \\ &= 2(2v_{p'h',ph} - v_{p'h,ph'}) \end{aligned} \tag{5.6}$$

The random phase approximation [14,16,17] is a device for computing approximate excitation energies (ω_ν) and excited state vectors $|\nu>$ from the HF ground state. Here

$$|\nu\rangle = Q_\nu^\dagger|\rangle, \quad Q_\nu^\dagger \equiv \sum_{ph}\sum(X^\nu_{ph}E_{ph} - Y^\nu_{ph}E_{hp}) \tag{5.7}$$

which by superoperator, equations of motion or time-dependent Hartree-Fock theories yield equations of the form of (5.5) with $K_{xx} = -K^*_{yy} = 2$. We have also developed the unitary group formulation of open-shell, restricted HF and RPA theories for all multiplicities [13,16]. Multi-configurational HF theory employs multiconfigurational states

$$|(G)\rangle = \sum_F |F\rangle\langle F|(G)\rangle \tag{5.8}$$

where $|F\rangle$ is a Gel'fand state. Following work of Dalgaard and Jørgensen [18] the variation function is

$$|(v)\rangle = e^S e^T|(0)\rangle \tag{5.9}$$

where $|(0)\rangle$ is the multiconfigurational ground state and $e^T(T \equiv \sum_{r>s} T_{rs}(E_{rs}-E_{sr}))$ and $e^S(S \equiv \sum_{(G)} S_{(G)}(|(G)\rangle\langle(0)| - |(0)\rangle\langle(G)|))$ are unitary transformations. By the BCH expansion and the variation of T and S to second order of the energy of the state (4.9) we obtain the equations shown in Table 3. (Their explicit form will be given elsewhere [19].) This second order, one step procedure has been shown to be quadratically convergent [20,21].

B. Effective Hamiltonians

We partition $V_F \equiv V_\rho^{[\lambda]}$ as follows:

$$V_F: \ \{|F\rangle\} = V_P: \ \{|P\rangle\} \oplus V_Q: \ \{|Q\rangle\} \tag{5.10}$$

with projectors $\mathcal{F}, \mathcal{P}$ and $\mathcal{Q}$. We wish to improve the limited CI calculation which diagonalized H in V_P by replacing it by an effective Hamiltonian

$$\hat{H}_P = \sum_{PP'}\sum \hat{H}_{PP'}|P\rangle\langle P'| \tag{5.11}$$

for which we propose three forms: <u>Bloch-Lindgren theory</u> [22,23]. We let $|K\rangle \in V_F$ be an exact eigenvector of H, $\mathcal{P}|K\rangle = |\kappa\rangle$ its projection in V_P and let $\Omega = \mathcal{P} + \sum_{(n)=1}\Omega^{(n)}$ be the perturbation expansion of the wave operator to obtain the equations listed in Table 4. We note that the Bloch-Lindgren Hamiltonian is non-hermitian. Brandow [24] has discussed hermitian effective Hamiltonians. We obtain a hermitian Hamiltonian by setting $\hat{H} = e^{-T}He^T(T = \sum_{PQ}\sum T_{PQ}(|P\rangle\langle Q| - |Q\rangle\langle P|))$ and carrying out the BCH expansion. This procedure yields simultaneous polynomials of infinite degree in T_{QP} which can be reduced by truncation. <u>Multiconfiguration Coupled-Cluster Theory</u> [25]. Here truncation to the jth degree is accomplished by truncation at the jth commutator and by excluding those $|Q\rangle$ which require higher excitations from V_P. <u>Primas-Klein Theory</u> [26,27]. Here we expand $T = \sum_{(n)=1} T^{(n)}$ in a perturbation expansion to obtain the equations shown in Table 5. These effective Hamiltonians together with the GUGA evaluation of $V_{FF'}$ should have application to large scale calculations in reducing the size of a secular equation for a given level of accuracy or for raising the level of accuracy for a given size secular equation. One example is to restrict the space via a frozen core approximation in which case the effective Hamiltonian becomes a valence electron Hamiltonian [28].

ACKNOWLEDGMENT

We gratefully acknowledge the assistance of Preston Kilman and Cheryl Dixon.

Table 3. Multiconfiguration Hartree Fock Theory

$$|(v)\rangle = e^{T}e^{S}|(0)\rangle,$$

$$T=\sum_{r>s} T_{rs}(E_{rs}-E_{sr}) \quad S=\sum_{(G)} S_{(G)}(|(G)\rangle\langle(0)|-|(0)\rangle\langle(G)|)$$

where

$$\begin{pmatrix} T_{rs} \\ S_{(G)} \end{pmatrix} = \begin{pmatrix} (A-B)_{r's',rs} & (A-B)_{r's',(G)} \\ (A-B)_{(G'),rs} & (A-B)_{(G'),(G)} \end{pmatrix}^{-1} \begin{pmatrix} W_{r's'} \\ V_{(G')} \end{pmatrix}$$

$$(A-B)_{r's',rs}=\langle(0)|[E_{s'r'},H,(E_{rs}-E_{sr})]|(0)\rangle$$

$$(A-B)_{r's',(G)}=A^{\dagger}_{(G'),rs}-B^{T}_{(G'),rs}$$
$$=\langle(0)|[[E_{s'r'},H],(|(G)\rangle\langle(0)|-|(0)\rangle\langle(G)|]|(0)\rangle$$

$$(A-B)_{(G'),(G)}$$
$$=\langle(0)|[|(0)\rangle\langle(G')|,H,(|(G)\rangle\langle(0)|-|(0)\rangle\langle(G)|)]|(0)\rangle$$

$$W_{r's'}=\langle(0)|[E_{s'r'},H]|(0)\rangle$$

$$V_{(G')}=\langle(0)|[|(0)\rangle\langle(G')|,H]|(0)\rangle$$

Table 4. Bloch-Lindgren Theory

$$\hat{H}_{PP'}=\delta(P,P')E_P^o+\sum_{(n)}\sum_Q V_{PQ}\Omega_{QP'}^{(n)}$$

where

$$-\Omega_{QP}^{(1)}=V_{QP}/(E_Q^o-E_P^o)$$

$$-\Omega_{QP}^{(2)}=(\sum_{Q'}V_{QQ'}\Omega_{Q'P}^{(1)}-\sum_{P'}\Omega_{QP'}^{(1)}V_{P'P})/(E_Q^o-E_P^o)$$

$$-\Omega_{QP}^{(3)}=(\sum_{Q'}V_{QQ'}\Omega_{Q'P}^{(2)}-\sum_{P'}\Omega_{QP'}^{(2)}V_{P'P}-\sum_{P'}\sum_{Q'}\Omega_{QP'}^{(1)}V_{P'Q'}\Omega_{Q'P}^{(1)})/(E_Q^o-E_P^o)$$

Etc.

Table 5. Primas-Klein Theory

1. $\hat{H}^{(0)}=H^{(0)}$ Expansion

$$\hat{H}^{(1)}=[H^{(0)},T^{(1)}]+V$$

$$\hat{H}^{(2)}=[H^{(0)},T^{(2)}]+[V,T^{(1)}]+\tfrac{1}{2}[[H^0,T^{(1)}],T^{(1)}]$$

$$\hat{H}^{(3)}=[H^0,T^{(3)}]+[V,T^{(2)}]+\tfrac{1}{2}[[H^{(0)},T^{(2)}],T^{(1)}]$$
$$+\tfrac{1}{2}[[H^{(0)},T^{(1)}],T^{(2)}]+\tfrac{1}{6}[[[H^{(0)},T^{(1)}],T^{(1)}],T^{(1)}]$$

2. $\hat{H}_{PP'}=\sum_{(n)}\hat{H}_{PP'}^{(n)}$ Projection

$$\hat{H}_{PP'}^{(0)}=\delta(P,P')E_P^o$$

$$\hat{H}_{PP'}^{(1)}=V_{PP'}$$

$$\hat{H}_{PP'}^{(2)}=\delta(P,P')\sum_Q T_{QP}^{(1)}V_{PQ}+\sum_Q T_{QP}^{(1)}V_{QP'}+\tfrac{1}{2}\sum_Q T_{QP}^{(1)}T_{QP'}^{(1)}(E_Q^o-\ {}_P^o)$$
$$+\delta(P,P')\tfrac{1}{2}\sum_Q T_{QP}^{(1)}T_{QP}^{(1)}(E_Q^0-E_P^o)$$

$$\hat{H}_{PP'}^{(3)}=\delta(P,P')\sum_Q T_{QP}^{(2)}V_{PQ}+\sum_Q T_{QP}^{(2)}V_{QP'}+\tfrac{1}{2}\sum_Q T_{QP}^{(2)}T_{QP'}^{(1)}(E_Q^o-E_P^o)$$
$$+\tfrac{1}{2}\sum_Q T_{QP}^{(1)}T_{QP'}^{(2)}(E_Q^o-E_P^o)+\delta(P,P')\sum_Q T_{QP}^{(2)}T_{QP}^{(1)}(E_Q^o-E_P^o)$$

where $[\hat{H},P]=0$

and where

$$T_{QP}^{(1)} = \frac{-V_{QP}}{(E_Q^o - E_P^o)} \quad , \quad T_{QP}^{(2)} = \frac{(V_{PP} - V_{QQ})}{(E_Q^o - E_P^o)} T_{QP}^{(1)}$$

$$T_{QP}^{(3)} = \left(\frac{V_{PP} - V_{QQ}}{E_Q^o - E_P^o}\right) T_{QP}^{(2)} + \frac{1}{3}[T_{QP}^{(1)}]^3 + \frac{1}{6}\sum_{P'} T_{QP}^{(1)} [T_{QP'}^{(1)}]^2 + \frac{1}{6}\sum_{Q'} T_{QP}^{(1)} [T_{Q'P}^{(1)}]^2$$

REFERENCES

[1] Matsen, F. A.: 1974, Int. J. Quantum Chem. S8, pp. 379-388. See also: Harter, W. G.: 1973, Phys. Rev. A 8, pp. 2819-2827; Harter, W. G. and Patterson, C. W.: 1976, Phys. Rev. A13, pp. 1067-1076.

[2] Gel'fand, I. M., and Graev, M. I.: 1967, Am. Math. Soc. Transl. 64, pp. 116-216.

[3] Moshinsky, M.: 1968, "Group Theory and the Many-Body Problem," Gordon and Breach, New York.

[4] Biedenharn, L. C.: 1963, J. Math. Phys. 4, pp. 436-445.

[5] Paldus, J.: 1976, Theo. Chem.: Adv. Perspect. 2, pp. 131-290.

[6] Shavitt, I.: 1978, Int. J. Quantum Chem. S12, pp. 5-32.
: 1979, unpublished.

[7] Matsen, F. A.: 1978, Adv. in Quantum Chem. 2, pp. 223-250.

[8] Matsen, F. A.: 1964, Adv. in Quantum Chem. 1, pp. 59-114.

[9] Matsen, F. A.: 1976, Int. J. Quantum Chem. 10, pp. 525-542.

[10] Gilmore, R.: 1974, "Lie Groups, Lie Algebras, and Some of their Applications," Wiley, New York.

[11] Wybourne, B.: 1970, "Symmetry Principles and Atomic Spectroscopy," Wiley, New York.

[12] Brooks, B. R. and Schaefer, H. F.: 1979, J. Chem. Phys. 70, pp. 5092-5106.

[13] Matsen, F. A. and Nelin, C. J.: 1979, Int. J. Quantum Chem. 15, pp. 751-767.

[14] Rowe, D. J.: 1970, "Nuclear Collective Motion," Methuen, London.

[15] Thouless, D. J.: 1961, "The Quantum Mechanics of Many-Body Systems," Academic, New York.

[16] Nelin, C. J.: 1979, Int. J. Quantum Chem., (in press).

[17] See for example: Jørgensen, P.: 1975, Ann. Rev. Phys. Chem. 26, pp. 359-380; Dunning, T. and McKoy, V.: 1967, J. Chem. Phys. 47, pp. 1735-1747.

[18] Dalgaard, E. and Jørgensen, P.: 1978, J. Chem. Phys. 69, pp. 3833-3844.

[19] Matsen, F. A. and Nelin, C. J.,: 1979, Int. J. Quantum Chem. (to be submitted).

[20] Yeager, D. L. and Jørgensen, P.: 1979, J. Chem. Phys. 71, pp. 755-760.

[21] Roothaan, C. C. J., Detrich, J. and Hopper, D. G.: 1979, Int. J. Quantum Chem. (in press).

[22] Bloch, C.: 1958, Nucl. Phys. 6, pp. 329-352.
[23] Lindgren, I.: 1978, Int. J. Quantum Chem. S12, pp. 33-57.
[24] Brandow, B. H.: 1979, Int. J. Quantum Chem. 15, pp. 207-242.
[25] Matsen, F. A.: 1979, Int. J. Quantum Chem. (in press).
[26] Primas, H.: 1963, Rev. Mod. Phys. 35, pp. 710-714.
[27] Klein, D. J. (private communication).

SYMPOSIUM II. BOND FORMATION AND BREAKING

Chairman : M. Simonetta

Instituto di Chimica Fisica
University of Milan, Milan, Italy

BOND FORMATION AND BREAKING: THE HEART OF CHEMISTRY

Massimo SIMONETTA

Istituto di Chimica Fisica e Centro CNR,
via Golgi 19, 20133 Milano, Italy

Theoretical approaches to the study of bond formation and breaking are briefly reviewed. The sensitivity of such processes to environmental effects is stressed.

It is the chemist's belief that molecules are made of atoms. Atoms in a molecule may be bonded or non bonded, and a molecule is described by defining both the component atoms and the bonds between them. Chemistry is the art of transforming molecules in other molecules, and this transformation is achieved by breaking and/or forming bonds. So it is no surprise that theoretical chemistry has devoted so many efforts to the elucidation of such processes. One problem arising here is the variation of bond distances, so that geometries substantially different from equilibrium are to be considered. The usual SCF approximation is not applicable and configurational mixing becomes essential. Special techniques to evaluate the optimized functions are available and can be applied to determine energy barriers for chemical reactions (1). A fundamental improvement in our understanding of molecular structure and its change when reaction occurs can be obtained by the study of the quantum topology of the molecular charge distribution characterized in terms of the vector field of charge density. Wave functions of nearly Hartree-Fock quality are in general sufficient here, but in a few examples configurational mixing is again needed (2,3). Molecular beams experiments made it possible to observe reactive atomic and molecular collisions and to extract informations on intermolecular potentials. In simple situations, e.g. the $F + H_2 \rightarrow FH + H$ reaction, quantum mechanical scattering theory can be applied and interesting features such as total reaction cross sections as a function of energy or product state rotational energy distribution are calculated (4). In a reacting molecule one or more vibrational modes can degenerate

K. Fukui and B. Pullman (eds.), Horizons of Quantum Chemistry, 51–56.

into translation and reactivity can be looked at as the consequence. In this view high accuracy in the potential surfaces of molecules becomes compulsory and anharmonic contributions must be taken into account. Reliable potential surfaces for polyatomic molecules can only be obtained by combining experimental and theoretical results (5). Among important recent contributions to an understanding of chemical reactivity I cannot help mentioning the introduction of symmetry rules (6), that had a tremendous impact on the development of mechanistic organic chemistry in the last decade; the frontier orbital theory (7) developed in Japan by K.Fukui, and the investigations carried on by L.Salem on the destiny of electron pairs in chemical reactions (8).

All theories so far mentioned consider the transformation in a single isolated molecule or the result of collisions between two molecules. Chemistry in our laboratories and in industrial plants is usually carried on in completely different conditions and the influence of surrounding cannot be dismissed since it can have a prevalent role in determining the chemistry. A few examples can illustrate the situation. Methyl-p-dimethylaminobenzene sulfonate reacts in the solid state to give the zwitterion p-trimethylammonium benzene sulfonate, while it is indefinitely stable in solution at room temperature (9). The rate of solid state rearrangement increases with temperature, but when the melting point is reached a sharp decrease in the rate of conversion is observed. The solid state reaction is intermolecular. These facts are consistent with the crystal structure of the reactant, shown in Figure 1. The arrangement of molecules in the crystal is

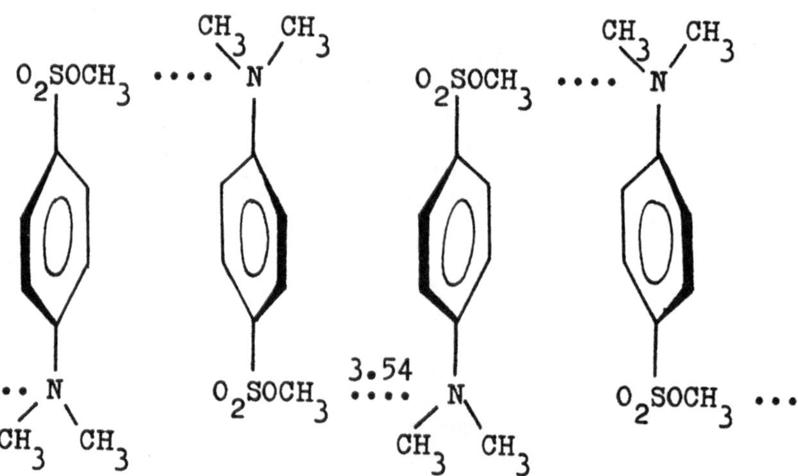

Figure 1. Arrangement of the molecules in the crystal structure of methyl-p-dimethylaminobenzene sulfonate.

such that migration of the methyl group from sulfonic group to the amino group of the neighbouring molecule is strongly favoured. Formally, the migrating group is the methyl cation CH_3^+, but this species in fact never forms since while the group donates its electrons the oxygen atom, it attracts the lone pair electrons of the nitrogen. Empirical molecular orbital calculations show that the reaction proceeds in two steps, with formation of an intermediate ion pair. An intermediate was in fact found but not identified in the experimental work. Force field calculations have shown that intermolecular interactions, as obtained in the Kitaigorodsky's atom pair approximation, increase only by a few per cent the energy barrier for the reaction and the enhancement of the reaction in the crystal can be rationalized as a consequence of a favourable entropy factor (10).

A second example is the dediazoniation of benzenediazonium ions in homogeneous solution (11). It has been found that dissociation of the cation occurs through the formation of at least one intermediate, an ion-molecule pair. In fact the reaction is in competition with N_α - N_β rearrangement and, when carried out in the presence of N_2 at 300 atm, with the exchange reaction. Kinetic data also support this mechanism. The energy variation along the dissociation path has been obtained by an empirical molecular orbital calculation, both for the isolated cation and for a supermolecule in which the cation is surrounded by the first shell of solvent (water) molecules. The results are shown in Figure 2. The

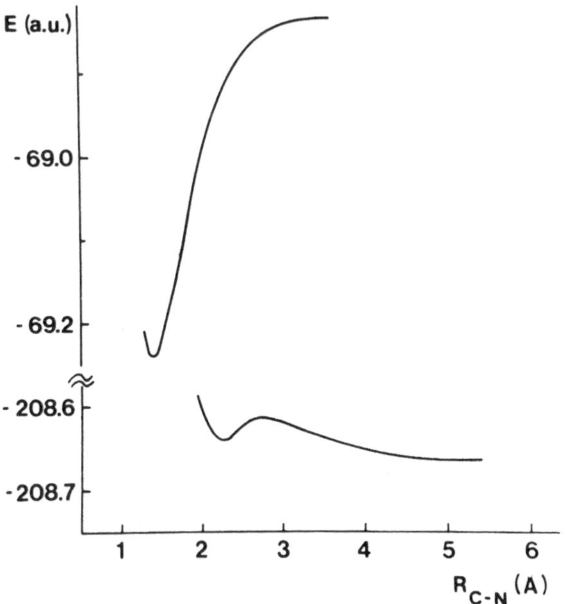

Figure 2. Dissociation energy in the gas phase (upper curve) and in solution (lower curve).

existence of the ion molecule pairs in solution is clearly indicated. The water molecules form a single cage around the pair, so that this can be identified as a "tight" ion-molecule pair (12). The results are in line with the experimental work of Saul Winstein on the formation of ion pairs in solvolysis of alkyl halides (13).

As a final example let us consider the rearrangement and hydrogenation of acetylene on a Pt(111) surface. Acetylene is adsorbed on such a surface at low temperature. At room temperature however a transformation is observed and the organic adsorbate has been identified as vinylidene, ethylidyne or ethylidene (14), when different techniques were used for the identification: ultraviolet photoelectron spectroscopy, low energy electron diffraction, or high resolution electron energy loss spectroscopy, respectively. The transformation of acetylene into vinylidene in the gas phase, and into the same product or CCH_3 or $CHCH_3$ on a cluster of Pt atoms simulating the (111) surface has been studied, again by empirical MO methods (15). The results are shown in Figure 3. It appears that while in the gas phase

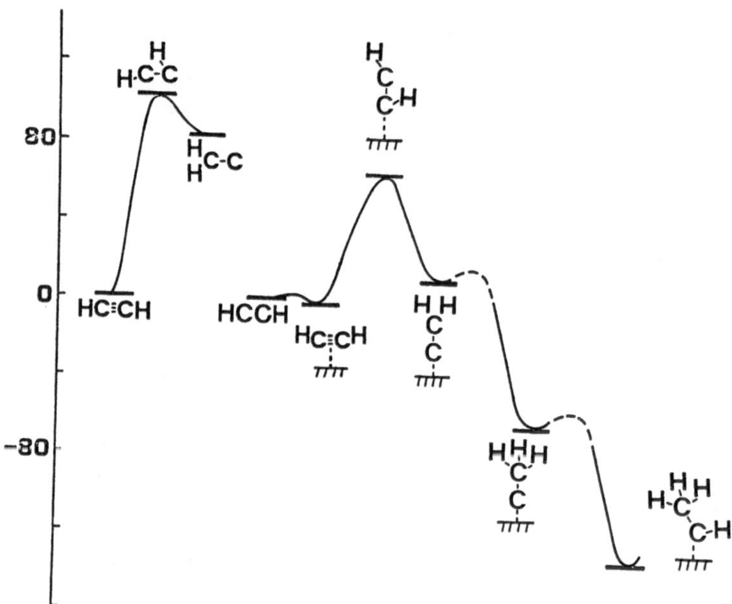

Figure 3. Energetics of the rearrangement and hydrogenation of acetylene in the gas phase and on a Pt(111) surface.

acetylene is strongly favoured in stability with respect to vinylidene, the two species have about the same energy when adsorbed on the metal surface. The hydrogenation of CCH_2 to CCH_3 and $CHCH_3$ occurs with no activation barrier if atomic hydrogen is available. The barrier seen on the figure is due to desorption of atomic hydrogen from the surface. It appears that the nature of the adsorbed species at room temperature is dependent on the availability of atomic hydrogen in the system; different results with different techniques can be justified. Acetylene is found to be adsorbed parallel to the surface on top of a metal atom, the other species seat perpendicularly on a triangular site, in agreement with LEED results. The catalytic effect of Pt on the rearrangement is also evident.

These few examples show convincingly that bond formation and bond breaking are extremely sensitive to environmental effects and theory should be able to deal with real situations rather than with idealized systems. It is a goal of quantum chemistry at present time to translate the above qualitative speculations into quantitative theoretical results.

REFERENCES

(1) K.Ruedenberg, Third International Congress of Quantum Chemistry, Kyoto, Japan 1979.

(2) R.F.W.Bader, S.G.Anderson, and A.J.Duke, J.Amer.Chem. Soc. 1979, 100 p.1389.

(3) R.F.W.Bader,T.Tung Nguyen-Dang, and Yoram Tal, J.Chem. Phys. 1979, 70 p.4316.

(4) J.S.Hutchinson, and R.E.Wyatt, J.Chem.Phys. 1979, 70 p.3509.

(5) P.Pulay, Third International Congress of Quantum Chemistry, Kyoto, Japan 1979.

(6) R.B.Woodward, and R.Hoffmann, The Conservation of Orbital Symmetry, Verlag Chemie, Weinheim 1970.

(7) K.Fukui, Accts.Chem.Res. 1971, 4 p.57.

(8) L.Salem, N.Journal de Chimie 1978, 2 p.529.

(9) C.N.Sukenik, J.A.P.Bonapace, N.S.Mandel, P.Lau, G.Wood, and R.G.Bergman, J.Amer.Chem.Soc. 1977, 99 p.851.

(10) A.Gavezzotti, and M.Simonetta, N.Journal de Chimie 1978, 2 p.69.

(11) J.Szele, and H.Zollinger, J.Amer.Chem.Soc. 1978, 100 p.2811.

(12) A.Gamba, M.Simonetta,G.Suffritti, J.Szele, and H. Zollinger, J.Chem.Soc. Perkin II 1979, in the press.

(13) S.Winstein, in Chimica Teorica, 1965, Acc.Naz.Lincei, Roma, p.307.

(14) L.L.Kesmodel, L.H.Dubois, and G.Somorjai, J.Chem.Phys. 1979, 70 p.2180.

(15) A.Gavezzotti, and M.Simonetta, Surface Science 1979, submitted.

QUANTUM TOPOLOGY

R.F.W. Bader
Department of Chemistry
McMaster University
Hamilton, Ontario L8S 4M1
Canada

ABSTRACT

This paper reviews a new approach to the problem of molecular structure and its morphogenesis: the theory of quantum topology. The observational basis for this theory is found in the topological properties of the total charge distribution of a molecular system. The quantum mechanical basis is provided by a generalization of the variational principle which defines and describes the properties of a subspace of a total system. Both the quantum and topological approaches define an atom in a molecular system, and the two definitions are equivalent.

Nowhere is the concept of structure more important than it is to the study of chemistry. The notion of a molecule being a collection of atoms joined by some network of bonds, i.e., the notion of molecular structure (as distinguished from the geometrical structure as determined by the relative positions of the nuclei in a molecule) plays an essential role in our understanding of chemistry. Due in large part to the intuitive ideas of Lewis, Pauling and Slater the concept of molecular structure evolved to its present point of operational usefulness notwithstanding the fact that it had never been formalized. From an extensive study of the properties of molecular charge distributions and their quantum mechanical basis, a new theory has been developed which both defines and provides a physical understanding of molecular structure and the related property of structural stability.

The primary concepts of molecular structure, those without which there would be no correlation and no prediction of the observations of descriptive chemistry are (1) the existence of atoms in molecules, (2) the ability to identify an atom (or a functional grouping of atoms) in a molecule by its characteristic properties, and (3) the concept of bonding - that molecular stability may be understood by assuming the existence of particularly strong interactions - bonded interactions - between certain pairs of atoms within the molecule. We distinguish between these primary concepts which have evolved as a direct result of our attempts

K. Fukui and B. Pullman (eds.), Horizons of Quantum Chemistry, 57–61.

to understand chemical observations, and the models which have evolved to rationalize these concepts. The concepts we hope to discover intact in the properties of the charge density; the models may survive totally, partially, or not at all.

It is our thesis that these primary concepts have meaning in real Euclidean space and hence they should be defined by properties of the system which are themselves manifest in real space - the properties of the charge density. This thesis may be demonstrated by first determining the universal properties of molecular charge distributions and then establishing the existence of a mapping of the experimentally determined concepts of chemistry onto this set of universal properties. The resulting theory, called quantum topology /1-4/, has recently been reviewed /5/. We give here only a brief summary of its basis and results.

The chemical concepts of atoms, bonds, structure and structural stability find precise physical expression in terms of the topological properties of the electronic charge density, $\rho(\mathbf{r},\mathbf{X})$, as displayed in the behaviour space $\mathbf{r}\varepsilon R^3$ and the change in these properties as the set of nuclear position coordinates $\mathbf{X}\varepsilon R^N$ of the control space is varied. The study proceeds by first determining the set theoretic properties of the collection of gradient paths g_m through $\rho(\mathbf{r},\mathbf{X})$, the trajectories of $\nabla\rho(\mathbf{r},\mathbf{X})$, associated with each critical point in $\rho(\mathbf{r},\mathbf{X})$, a point where $\nabla\rho(\mathbf{r},\mathbf{X})=\mathbf{0}$. The essential observation is that the only local maxima of a ground state charge distribution occur at the positions of the nuclei. (On the basis of many studies conducted in this laboratory, this same observation would seem to apply to excited states of many-electron systems as well.) The nuclei are, therefore, identified as point attractors of the gradient vector field $\nabla\rho(\mathbf{r},\mathbf{X})$ of the charge density, an attractor being a particular closed subset $G\subset R^3$ of the g_m. In particular, every g_m originating in G, (a g_m whose α-limit set contains a point in G), is itself contained in G. For each attractor there exists an open invariant neighbourhood B of G, such that every trajectory starting in B has G as its terminus (ω-limit set). The associated basins partition a molecular system into atomic fragments. <u>An atom is defined as the union of an attractor and its basin</u>. Each atom is a stable structural unit (stable in terms of a local homeomorphism $h:R^3\rightarrow R^3$ such that h maps trajectories of a given $\nabla\rho(\mathbf{r},\mathbf{X})$ onto trajectories of $\nabla\rho(\mathbf{r},\mathbf{X}')$ where X' is sufficiently close to $\mathbf{X}$). Hence an atom is uniquely defined as the basic structural unit of matter by the topological properties of $\rho(\mathbf{r},\mathbf{X})$. The common boundary of two neighbouring atomic fragments, the interatomic surface, is defined by a closed subset of the g_m, all members of which have as their ω-limit set a particular critical point of $\nabla\rho(\mathbf{r},\mathbf{X})$. Such a critical point also generates a pair of gradient paths linking the two neighbouring attractors. The union of this pair of gradient paths and their end points is called a <u>bond path</u>. The network of bond paths defines a <u>molecular graph</u> of the system. It has been demonstrated that this network coincides with the one generated by linking together those pairs of atoms which are considered to be bonded to one-another on the basis of chemical observations.

Having defined a unique molecular graph for any molecular geometry, the total nuclear configuration space is partitioned into a number of regions. Each region is associated with a particular structure defined as an equivalence class of molecular graphs. By its definition as an equivalence class, the interior of a given structural region is an open set and points in such a region of configuration are, therefore, structurally stable. The boundary separating two structural regions is, however, a closed set. Points on the boundary denote unstable structures and they represent the loci of chemical change. A chemical reaction in which chemical bonds are broken and/or formed is, therefore, a trajectory in configuration space which must cross one of the boundaries between two neighbouring structural regions. These boundaries form the catastrophe set of the system. A knowledge of the catastrophe set enables one to construct a structure diagram for a molecular system which, like a phase diagram in thermodynamics, denotes the boundaries separating regions of different structure (phase) as a function of the control parameters. A structure diagram embodies a knowledge of the totality of all possible structures for a given system and of all possible mechanisms for structural change within the system.

It has been demonstrated that these definitions of structure and structural stability persist beyond the adiabatic approximation. Quantum mechanically, a structure is associated with an open neighbourhood of the most probable nuclear geometry.

We have previously demonstrated /6/ that the topological partitioning of a molecule into atoms coincides with the quantum mechanical partitioning of a molecule into subspaces with particular quantum mechanical properties: both partitionings are defined by the unique set of surfaces in a molecule (or a solid) which satisfy the condition of zero flux in $\nabla\rho(\underset{\sim}{r})$,

$$\nabla\rho(\underset{\sim}{r})\cdot\underset{\sim}{n} = 0 \qquad \forall\underset{\sim}{r}\in S(\underset{\sim}{r}) \tag{1}$$

The quantum mechanical basis for the definition of an atom and its properties has been extended /3,4/. The original generalization of the quantum mechanical variational principle to any region Ω bounded by a surface of zero flux

$$\delta G'(\hat{A}\psi,\Omega) = \tfrac{1}{2}\,\varepsilon\{<[\hat{H},\hat{A}]>_{\Omega} + c.c.\}/<\psi,\psi>_{\Omega} \tag{2}$$

(where $G(\psi,\Omega)$ is the energy of the total system expressed as a functional of ψ and $\delta\psi=\varepsilon\hat{A}\psi$) has been extended to the general time dependent case to obtain

$$\delta L(\hat{A}\psi,\Omega) = -\tfrac{1}{2}\varepsilon\{<[\hat{H},\hat{A}]>_{\Omega} + c.c.\}/<\psi,\psi>_{\Omega} \tag{3}$$

where $L(\psi,\nabla\psi,\psi,t)$ is the Lagrangian of the system. This work led in a natural way to a generalization of Schwinger's quantum action principle to obtain the quantum definition of an atom and its properties. This principle states that the general variation of the action integral operator

$$\hat{W} = \int_{t_1}^{t_2} \hat{L}(\psi,\nabla\psi,\dot{\psi},t)\, dt \qquad (4)$$

is given by the difference in the values of the generators of the unitary transformations causing the change in the quantized system at the two time end points. Through this principle one is able to obtain a description of an atom's properties in terms of the formalism of infinitesimal unitary transformations. In the wave function picture we obtain Ehrenfest equations of motion for a property of an atom in a molecule and relate these to a variation in the total energy. Using the quantum action principle, we obtain Heisenberg equations of motion for atomic operators and relate these to changes in the action over an atom in a molecule. These considerations lead to a general quantum mechanical definition of an atom (free or bound) as a region of space Ω for which the integral

$$(\hbar^2/4m) \int_{t_1}^{t_2} dt \, \{\int_\Omega \nabla^2\rho(\underset{\sim}{r})\, d\underset{\sim}{r}\} \qquad (5)$$

makes no contributions to the action for finite or infinitesimal changes in the system. This definition is necessarily satisfied if the boundary of the atom at all times satisfies the topological requirement that it possess a zero flux in $\nabla\rho(\underset{\sim}{r})$. In summary, we have a quantum topological definition of an atom and of its properties.

By their very definition, the topologically defined atoms are the most transferable fragments of a molecular system. From the quantum mechanical description of these atoms one obtains both an explanation and prediction of the existence of additivity schemes in chemistry, and more importantly, of the experimental observation that atoms or polyatomic functional groups may exhibit characteristic sets of properties which vary between relatively narrow limits. The present theory demonstrates that the charge distribution of an atom is the ultimate carrier of chemical information.

The nuclear positional coordinates are the control parameters in the study of the topological properties of $\rho(\underset{\sim}{r},\underset{\sim}{X})$. From our studies it is clear that the nuclear potential plays the dominant role in determining the topological properties of $\rho(\underset{\sim}{r},\underset{\sim}{X})$. To investigate this relationship between $V(\underset{\sim}{r},\underset{\sim}{X}) = -\Sigma Z_\alpha (|\underset{\sim}{r} - \underset{\sim}{X}_\alpha|)^{-1}$ and $\rho(\underset{\sim}{r},\underset{\sim}{X})$ in a more thorough manner we have investigated the topological structure of $V(\underset{\sim}{r},\underset{\sim}{X})$ through a study of its associated gradient vector field $\nabla V(\underset{\sim}{r},\underset{\sim}{X})$ /7/. We are able to present strong evidence that in general, the structures of the $\rho(\underset{\sim}{r},\underset{\sim}{X})$ and $V(\underset{\sim}{r},\underset{\sim}{X})$ fields will be homeomorphic for a given system. In principle this result implies that all of the information in the structure diagram for a system is determined by $V(\underset{\sim}{r},\underset{\sim}{X})$. The corresponding diagram for any particular quantum state as determined by the associated $\rho(\underset{\sim}{r},\underset{\sim}{X})$ is obtainable by the appropriate homeomorphic mapping of the structure diagram observed for $V(\underset{\sim}{r},\underset{\sim}{X})$.

An interesting problem is to find what connection exists between the topological instabilities in a system and the energetics of the system. On the purely observational level we have found (within computational accurary) that the topologically unstable structures (corresponding to a switching of attractors), in the isomerization reactions of HCN and CH_3CN to their corresponding isocyanides, occur at the nuclear configurations possessing the maximum energy along the minimum energy path: the catastrophe points are the transition states of these two reactions.

REFERENCES

1. R.F.W. Bader, S.G. Anderson and A.J. Duke: 1979, J. Am. Chem. Soc. 101, pp. 1389-1395.
2. R.F.W. Bader, T.T. Nguyen-Dang and Y. Tal: 1979, J. Chem. Phys. 70, pp. 4316-4329.
3. S. Srebrenik, R.F.W. Bader and T.T. Nguyen-Dang: 1978, J. Chem. Phys. 68, pp. 3667-3679.
4. R.F.W. Bader, S. Srebrenik and T.T. Nguyen-Dang: 1978, J. Chem. Phys. 68, pp. 3680-3691.
5. R.F.W. Bader, Y. Tal, S.G. Anderson and T.T. Nguyen-Dang: Israel J. Chem. (to appear in December, 1979).
6. S. Srebrenik and R.F.W. Bader: 1975, J. Chem. Phys. 63, pp. 3945-3961.
7. Y. Tal, R.F.W. Bader and J. Erkku: Phys. Rev. A (to appear in 1980).

QUANTUM MECHANICAL STUDY OF CHEMICAL REACTION DYNAMICS*

Robert E. Wyatt
The University of Texas
Austin, Texas 78712, USA

ABSTRACT

The quantum mechanical study of low energy atom-diatomic molecule collisions is reviewed. Following a brief introduction to the theory of electronic adiabatic collisions, reaction probabilities and cross sections for the three-dimensional $F + H_2$ reaction are discussed. These results indicate the participation of an internal excitation resonance in the reaction leading to HF(v=2) + H. The mechanism of resonant state formation in this reaction is explored in more detail with the presentation of a number of dynamic details of the collinear reaction. Attention is then shifted to electronic nonadiabatic reactions. A theory of collisions of the type $X(^2P) + H_2$, leading to electronic-rotational energy transfer, is presented. The theory is applied to spin-orbit quenching of $F(^2P)$ by H_2. Finally, some directions for future studies are indicated.

I. INTRODUCTION

The quantum mechanical study of chemical reaction dynamics has become incredibly more sophisticated during the past decade. Although these are certainly not routine calculations at this time, a number of atom plus diatomic molecule reactions will be extensively studied in the following decade. Some of the developments in this area which occurred during the 1970's (most of which concerned the hydrogen isotope reactions[1-3]) have been reviewed elsewhere.[4-8] We will be concerned here with a presentation and discussion of selected results (reaction probabilities, cross sections, and transition state dynamic details) for the $F + H_2 \rightarrow FH + H$ reaction.

*Supported in part by research grants from the Robert A. Welch Foundation and the National Science Foundation.

K. Fukui and B. Pullman (eds.), Horizons of Quantum Chemistry, 63–85.

In Section II, the electronic adiabatic $F + H_2$ reaction is considered in some detail. After briefly reviewing a quantum mechanical theory of three-dimensional reactions, computational results on resonance structure in the low energy reaction are described. Then, in order to understand the dynamic origin of this resonance, a number of details of the collinear reaction (local vibrational excitation probabilities, and scattering probability density and flux maps in the transition state region) are discussed. Section II then ends with new results on resonance structure in model chemical reactions.

In Section III, a theory is presented for electronic nonadiabatic processes which employs electronic information from diatomics-in-molecules (DIM) theory. This section ends with numerical results on quenching probabilities in $F(^2P) + H_2(j=0)$ collisions. Section IV concludes with a discussion of trends that may occur in future studies of atom-diatom chemical reactions.

II. ELECTRONIC ADIABATIC REACTIONS

A. Theory

As mentioned in the introduction, the theory of three-dimensional atom-diatom chemical reactions has advanced significantly during the past decade. The first quantum results beyond the hydrogen exchange reaction have been reported. These numerical results[9,10] for the $F + H_2 \rightarrow FH + H$ reaction will be described in later sections. However, in this section, a brief outline of quantum reactive scattering theory will be presented. Other recent reviews should be consulted for details.[7,1]

Step 1. Formulate Coordinates. Six coordinates are required to define the instantaneous size and shape of the three particle (nuclear) triange, in the CM frame.

a. Space Fixed to Body Fixed Coordinates. These six coordinates may be chosen to be the three Euler angles (θ,ϕ,χ) to orient the nuclear triangle relative to nonrotating xyz axes, plus three body fixed (BF) coordinates to define the size and shape of the triangle.

b. Body-Fixed Coordinates to Natural Collision Coordinates (NCC). Two distances and an angle are used as the BF coordinates. Natural collision coordinates[11] (s = translational coordinate, ρ = vibrational coordinate, γ = bending coordinate) are one such set.

c. Arrangement Tube Geometry. Three tubes leading from AB + C to A + BC and B + AC products and joining on arrangement tube matching surfaces are constructed in a coordinate space where collinear reactions occur in the xy plane, and deviation from collinearity is plotted along the positive z-axis. Figure 1 schematically shows three arrangement tubes for an atom-diatom reaction.

d. Matching Surfaces. Coordinates $(\theta\phi\chi s\rho\gamma)$ are defined on each tube, with $s = \pm\infty$ corresponding to separated reactants and products, and $s = 0$ defining a matching surface.

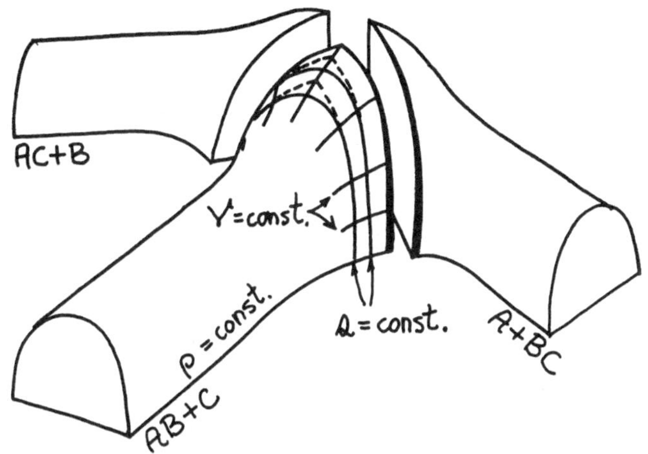

Figure 1. Arrangement tubes for the AB + C → A + BC, B + AC reaction. The tubes are drawn for one value of ρ, the vibrational displacement. On the reactant tube, curves of constant s and γ are drawn. The three tubes have been pulled apart slightly to show the matching surfaces.

Step 2. Hamiltonian Operator.

a. Kinetic Energy Operator. Although tedious to derive, the kinetic energy operator in NCC may be partitioned into rotational (triangle tumbling plus internal rotation-bending), vibrational and translational components, along with some coupling terms.

b. Potential Surface. For an electronic adiabatic reaction, the potential surface V(s, ρ,γ) may also be decomposed into translational, vibrational, bending, and vibration-bend coupling terms.

c. Partitioning. As a result of the partitioning of $\hat{T}$ and $\hat{V}$, the total Hamiltonian may also be decomposed into translational, vibrational, and rotational-bending components. The rotational Hamiltonian is that of a freely tumbling, but hindered asymmetric top, where the hindering arises from the angular anisotropy of the interaction potential, $V_{bend}(\gamma;s)$.

Step 3. Local Basis Sets.

a. Vibration. In order to build local adiabatic vibrational basis functions,

$$\left[\hat{T}_\rho + \hat{V}_{vib}(\rho;s)\right] H_v(\rho;s) = E_v(s)\, H_v(\rho;s) \quad (1)$$

the vibrational Hamiltonian may be diagonalized in a local harmonic basis, or if V_{vib} is a model (e.g. Morse) potential, then analytic vibrational functions are very useful.

b. Rotation. A useful basis for analyzing the rotational dynamics are the eigenstates of the hindered asymmetric top Hamiltonian,[12],

$$[\hat{T}_{rot} + \hat{V}_{bend}(\gamma; s)]\Omega^{JM}_{j\ell} = W^J_{j\ell}(s)\Omega^{JM}_{j\ell} \quad (2)$$

where $W^J_{j\ell}(s)$ is the local rotational energy (triangle tumbling plus hindered rotor energy), $j\ell$ are asymptotic rotational quantum numbers (j for molecular rotation and ℓ for atom-molecule orbital rotation), and $\Omega^{JM}_{j\ell}$ is the top function for total angular momentum J and space-fixed z-component M. The asymmetric top function is a sum of products over K (the component of J along the BF z-axis) of symmetric top functions N^J_{MK} times hindered internal rotor functions $R^{JK}_{j\ell}$.

c. Correlation Diagrams. Rotational correlation diagrams ($W^J_{j\ell}$ vs. s) and rovibrational correlation diagrams are useful in analyzing the subsequent dynamics since transitions frequently tend to take place near avoided crossings. Rotational energy correlation diagrams for the $H + H_2$ reaction[12] have been presented in detail.

Step 4. Close-Coupling Equations.

a. Wavefunction Expansion. At each value of s, the total scattering wavefunction, for a reaction initiated in channel $(vj\ell)$ may be expanded in products of local vibration-rotation basis functions:

$$\Psi^{JM}_{vj\ell}(\theta\phi\chi s\rho\gamma) = I^{-1/2}\sum_{v'j'\ell'} T^J_{v'j'\ell'vj\ell}(s)H_{v'}(\rho;s)\Omega^{JM}_{j'\ell'}(\theta\phi\chi\gamma;s) \quad (3)$$

where I is a scale factor to simplify the equations for the translational wavefunctions, $T^J_{v'j'\ell'vj\ell}(s)$.

b. Formulation of CC Equations. In order to derive coupled equations for the translational wavefunctions, we force $\Psi^{JM}_{vj\ell}$ to satisfy the Schrodinger equation, then project onto each basis state, one at a time:

$$\frac{d^2T^J_{v'j'\ell'vj\ell}}{ds^2} = \sum_{v''j''\ell''} V_{v''j''\ell''v'j'\ell'}T^J_{v''j''\ell''vj\ell} \quad (4)$$

These equations are conveniently solved via the R-matrix propagation[4,14] method, in which the integration range over s is broken into a sequence of small Δs segments over which the basis functions do not vary with respect to s. The equations to be solved within each segment are of the form

$$\langle H_{v'}\Omega^{JM}_{j'\ell'}|\hat{H}-E|\Psi^{JM}_{vj\ell}\rangle_{\theta\phi\chi\gamma\rho} = 0. \quad (5)$$

Solutions and their derivatives in different segments must satisfy continuity conditions at the segment boundaries. As a result, a set of N primative solutions are obtained, for N different input channels.

c. J_z-Conserving. Since many more vibration-rotation channels ($vj\ell$) may be open than the computer can accommodate, a rotational channel reduction scheme is required. In one version of the J_z-conserving[15] scheme for reactive collisions,[16] only those asymmetric top states $\Omega^{JM}_{j\ell}$ which correlate to specific bending states in the transition state region are employed. For reactions dominated by near-linear intermediates, asymptotic states correlating to K = 0 prolate symmetric top states in the s = 0 reactive region are selected.

Step 5. Boundary Conditions.

a. Matching Surface Continuity. The primative solutions on each arrangement tube are required to be continuous, with continuous first derivatives, across the matching surfaces which separate different arrangement tubes.

b. Asymptotic Boundary Conditions. As $|s| \to \infty$, the translational wavefunctions are required to satisfy the boundary conditions ($\alpha = vj\ell$)

$$\text{elastic channel:} \quad T^J_{\alpha\alpha}(\mathcal{R}) = k_\alpha^{-1/2}\left[e^{ik_\alpha \mathcal{R}} - S^{J(-)}_{\alpha\alpha} e^{-ik_\alpha \mathcal{R}}\right]$$

$$\text{reactive channel:} \quad T^J_{\alpha'\alpha}(\mathcal{R}) = -k_{\alpha'}^{-1/2} S^{J(+)}_{\alpha'\alpha} e^{+ik_{\alpha'}\mathcal{R}} \tag{6}$$

where the outgoing wave amplitudes are elements of the S-matrix. They are probability amplitudes for processes $\alpha \to \alpha'$ at total energy E.

Step 6. Physical Results.

a. Reaction Probabilities. From the S-matrix elements, the state-to-state reaction probability at total angular momentum J and energy E is

$$P^J_{vjv'j'}(E) = \sum_{\ell=|J-j|}^{J+j} \sum_{\ell'=|J-j'|}^{J+j'} \left|S^{J(+)}_{vj\ell v'j'\ell'}(E)\right|^2 \tag{7}$$

b. Cross Sections. The degeneracy averaged reaction cross section, in terms of reaction probabilities is

$$\sigma_{vjv'j'}(E) = \frac{\pi}{(2j+1)k_\alpha^2} \sum_{J=0,1..} (2J+1) P^J_{vjv'j'}(E) \tag{8}$$

The growth of σ near the reaction threshold (if there is one) and possible post threshold resonance structure, are particularly interesting features of the energy dependence of σ.

c. Dynamic Details. Once the translational wavefunctions are computed, the total wavefunction in the transition state region may be obtained. From the translational wavefunctions, the energy partitioning into various vibrotor basis states may be generated, and from the total wavefunction, the probability density and probability flux distribution may be generated.

The asymmetric top-natural coordinate method outlined above has been applied to the three-dimensional $H + H_2$ and $F + H_2$ reactions.[1,10]

Because of the complexity of the kinetic energy operator in NCC, a near linear intermediate approximation was employed: coupling terms which become important for large γ (bent transition state) in the region near s = 0 were neglected. We now proceed to computational results for the $F + H_2$ reaction.

B. The $F + H_2$ Reaction

Quantum close-coupling calculations for the collinear $F + H_2$ reaction clearly demonstrate low energy nonclassical resonance structure in the $v = 0 \rightarrow v' = 2$ reaction probability curve.[17-20] The origin of this resonance will be discussed in more detail in the next section. First, in this section, recent quantum calculations on the three-dimensional reaction[10] will be reviewed, and evidence for low energy resonance structure will be presented.

The total wavefunction at position s along the reaction coordinate and at each value of J and M is expanded in products of local asymmetric top functions times Morse oscillator functions (see Eq. (3)). For collisions in the energy range $0.05 \leq E_{tr} \leq 0.23$ eV. ($E_{total} = E_{tr} + 0.27$), 60 local rovibrational channels were included in the basis, with the following number of rotational functions (even j + odd j) in each vibrational manifold: (12, 12, 12, 8, 6, 2, 2, 2, 2, 2), in which the top five vibrational manifolds are closed in products at the highest energy studied. The J_z-conserving approximation was used at each J to select the rotational channels, for $0 \leq J \leq 26$. The Muckerman V collinear potential surface[22] was employed together with the approximate bending potential $V_{bend}(\gamma,s) = \frac{1}{2}v_0(s)[1-\cos 2\gamma]$. This potential is similar to one employed in recent semiclassical trajectory calculations.[21]

Numerical integration of the CC equations (Eq. (5)) yields elements of the S-matrix for the processes $(v=0, j=0, \ell) \rightarrow (v', j', \ell')$. Reaction probabilities at each J and E, defined by summing over all product rotational indices $(j'\ell')$,

$$P^J_{0v'}(E) = \sum_{j'=0}^{(open)} \sum_{\ell'=|J-j'|}^{J+j'} |S^J_{00J\rightarrow v'j'\ell'}(E)|^2 \tag{9}$$

are plotted in Fig. 2 for v' = 2 and 3. When J = 0, the rapid growth and then decline in P^0_{02} as E increases is qualitatively similar to the collinear reaction probability curve.[18,19] However, the shift in the peak probability to higher J (note that since j = 0, J = ℓ, the reactant orbital angular momentum quantum number, which is semi-classically related to the reactant impact parameter $\mu v b = [\ell(\ell+1)]^{\frac{1}{2}}\hbar$) as E increases leads to a "resonance ridge" in the reaction probability surface. By way of contrast, the $0 \rightarrow 3$ reaction probability surface shows a gradual decline as ℓ increases at each value of E. The slow growth of P^J_{03} as E increases, at J=0 for example, is again qualitatively similar to the collinear behavior.[18,19]

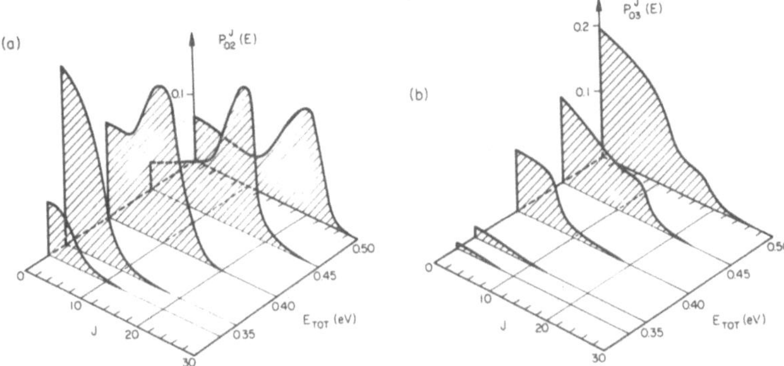

Figure 2. Reaction probability surfaces[10] for F + H_2(v=0,j=0) → FH(v',all j') + H for (a) v'=2, (b) v'=3. Translational energy = total energy - 0.27 eV.

Total reaction cross sections are obtained from the reaction probabilities by summing over J and M (M summation leads to the usual (2J + 1) factor), as in Eq. (8). The total cross sections in Fig. 3 show a broad 0 → 2 resonance and a slowly increasing 0 → 3 cross section. The 0 → 2 cross section is much wider than the collinear reaction probability due to the participation of many rotors, and to the detailed structure of the resonance ridge in Fig. 2a. In addition, the threshold for the 3D reaction is about 0.05 eV higher than the collinear result.

The shape of the reaction probability surface in Fig. 2a has very interesting implications for the angular distribution of the scattered products. Professor Lee[23] has pointed out that the sudden shift in the peak reactivity away from ℓ=0 as the translational energy increases by 3 kcal/mole between E_{tot} = 0.31 and 0.40 eV implies that the peak intensity on a velocity-scattering angle map should shift from back-scattering to sideways scattering (this is because increasing ℓ causes the product HF molecules to scatter in a more forward direction). Detailed crossed molecular beam experiments on the F + H_2(D_2) reactions are currently being analyzed and it appears that the data are in at least qualitative agreement with the above conjecture. These new experiments should provide more detail on the energy dependence of

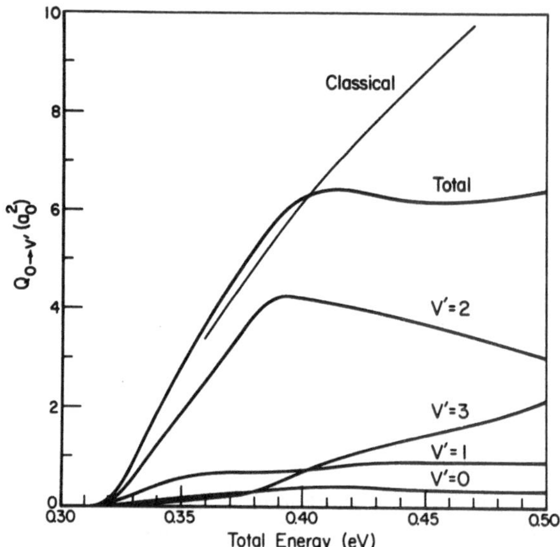

Figure 3. Total reaction cross sections[10] for the $F + H_2 \rightarrow FH(v') + H$ reaction. The total quantum mechanical cross section is compared to the quasiclassical[22] result on a slightly different potential surface.

these reactions than could be obtained in earlier pioneering studies.[24]

C. Collinear $F + H_2$ Resonances

Resonances are clearly a very important feature of the $F + H_2$ reaction. In order to understand their dynamical origin, we will examine quantum and classical results for the collinear FH_2 reaction.[25,26] First, the quantum results were obtained by expanding the wavefunction in a local vibrational basis,

$$\Psi_v(s,\rho) = (1+K\rho)^{1/2} \sum_{v'} T_{v'v}(s)\, H_{v'}(\rho;s) \tag{10}$$

Once more, s is the translational coordinate, ρ is the vibrational coordinate, and K is the reaction coordinate curvature). The translational wavefunctions $T_{v'v}(s)$ were found by numerically integrating the CC equations; usually 10 local Morse vibrational functions were used. For comparison, quasiclassical results were obtained by integrating ensembles of trajectories, starting with $H_2(v=0)$. Trajectories between s and s+Δs were analyzed to obtain vibrational populations, $P_v(s)$, by computing the local vibrational energy for each trajectory, and assigning the trajectory to a bin corresponding to the (nonintegral) quantum number.

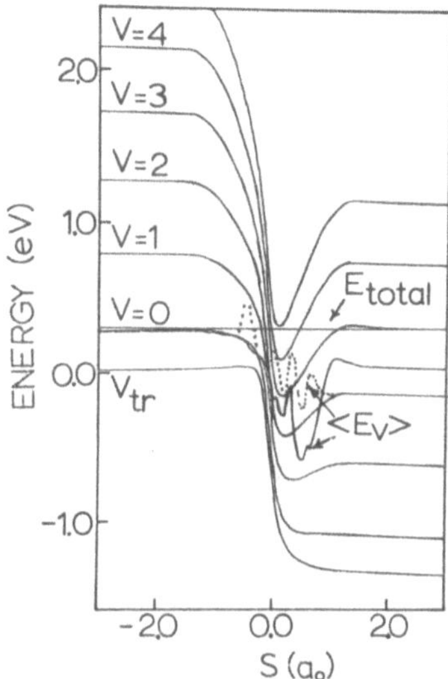

Figure 4. Local average vibrational energy, $<E_v>$, vs. reaction coordinate (s) for the collinear F + H_2 reaction: solid line, classical result[24] at 0.30 eV total energy; dotted line, quantum result[25] at 0.30 eV.

The first comparison, in Fig. 4, shows the classical and quantum average vibrational energies superimposed on the vibrational energy correlation diagram which links reactant H_2 states to product HF states. The classical results are at E_{total} = 0.30 eV total energy, slightly above the classical reaction threshold. The quantum results are at E_{total} = 0.284, the v=0 → v=2 collinear resonance energy. The classical and quantum $<E_v>$ are similar until the region between s = 0.0 to s = 1.5a_o, where the upper vibrational wells in the FHH region dip below their asymptotic HF values. In this region, the quantum values of $<E_v>$ are considerably higher than the classical values. Finally, on the approach into products, the classical value settles between the v=2 and v=3 HF vibrational energies. By way of contrast, the quantum reaction is much more specific, with $<E_v>$ settling neatly onto the v=2 HF curve.

The local vibrational probabilities in Fig. 5 reflect these differences. The classical and quantal probabilities in v=0 (Fig. 5a and Fig. 5d) both drop sharply near s = -0.1a_o. The v=1 level receives most of the excitation in the classical case (followed in importance by v=2

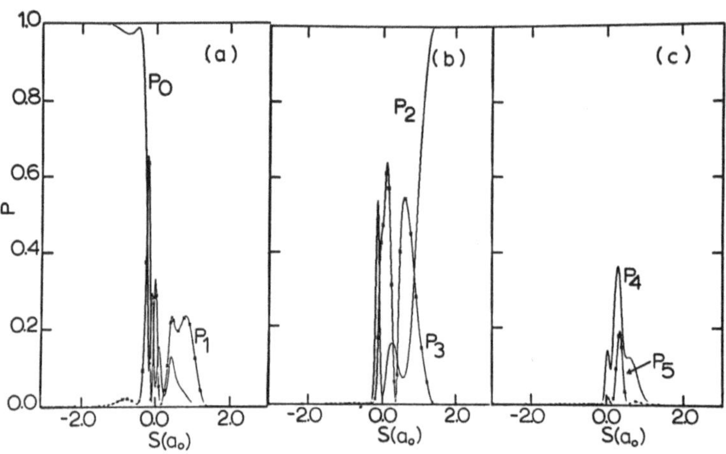

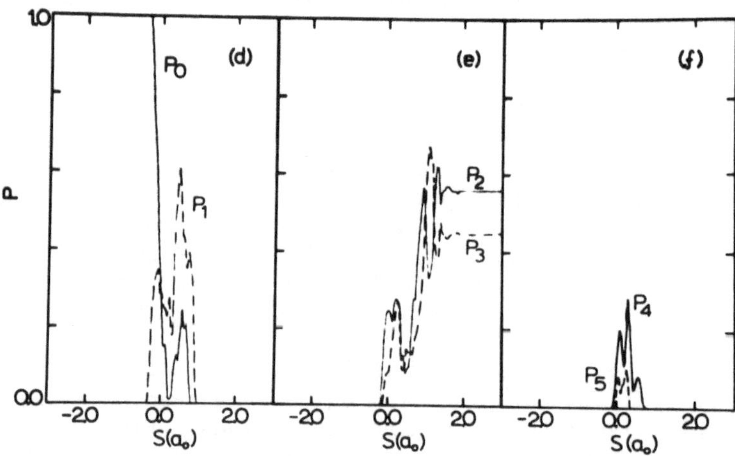

Figure 5. Local normalized vibrational state probabilities in the collinear $F + H_2$ reaction: (a)-(c) quantum results[26] at 0.284 eV total energy; (d)-(f) classical results[25] at 0.30 eV.

and v=3), but the quantum result is different; the excitation probabilities into v=2 and 3 are much higher than into v=1. Extensive excitation from v=0 into wells in the upper vibrational curves is the mechanism by which Feshbach (internal excitation) resonances are formed in this reaction. Finally, in late stages of the reaction ($s \gtrsim 1.5a_0$), the quantum probability in v=3 falls to zero, while the classical probability settles onto approximately 0.4. Thus, decay of the (quantum) resonant FHH complex is much more specific than can be accounted for by quasiclassical trajectories.

Another and possibly more effective way to display the nature of the low energy reactive resonance is to plot the local probability density and probability flux,[26] computed directly from the wavefunction in Ea. (10). These quantities are

$$P(\Omega,\rho) = |\Psi_v(\Omega,\rho)|^2,$$

$$\vec{F}(\Omega,\rho) = \frac{\hbar}{2\mu i}\left[\Psi_v^* \vec{\nabla}\Psi_v - \Psi_v \vec{\nabla}\Psi_v^*\right] \tag{11}$$

Fig. 6a shows a perspective plot of $P(s,\rho)$ near the interaction region, while Fig. 6b shows the flux map, looking down on the same region. The probability density shows strong interference effects in the reactant valley, due to elastically scattered H_2(v=0) interfering with incoming H_2(v=0). In the product valley, the emerging triple-peaked density corresponding to HF(v=2) is clearly seen. Most interesting, however, is the density buildup in the center of Fig. 6a; the three-peaked density <u>is</u> the resonant state. Slightly off resonance, the height of these peaks drops by about a factor of two. Not clearly shown in this figure are three nodes in the density which separate the four major peaks from one another. In Fig. 6b, the flux vectors are seen to whirlpool around each of these nodal points. These whirlpools are quantized[27] in the sense that the hydrodynamic circulation integral about each node is "quantized." If Ψ is written as $P^{1/2}e^{is/h}$, then the local flow velocity is $\vec{v} = \vec{F}/P = \mu^{-1}\vec{\nabla}s$. The circulation integral is a line integral along any closed contour surrounding the nodal point:

$$\oint \vec{v}\cdot d\vec{r} = \frac{\hbar}{\mu}\,\Delta\left(\frac{S}{\hbar}\right) \tag{12}$$

where $\Delta(S/\hbar)$ is the phase change of Ψ around the path. Continuity requires this to be $2\pi N(N = \pm 1, \pm 2 \cdots)$, so that the circulation is quantized:

$$\oint \vec{v}\cdot d\vec{r} = N\left(\frac{h}{\mu}\right) \tag{13}$$

For all energies which have been examined, $|N| = 1$, so these whirlpools are in their "ground state." The quantum nature of the F + H_2 reaction is clearly evident in the preceding figures. Regions of extensive excitation into high vibrational levels of the FH_2 complex are associated

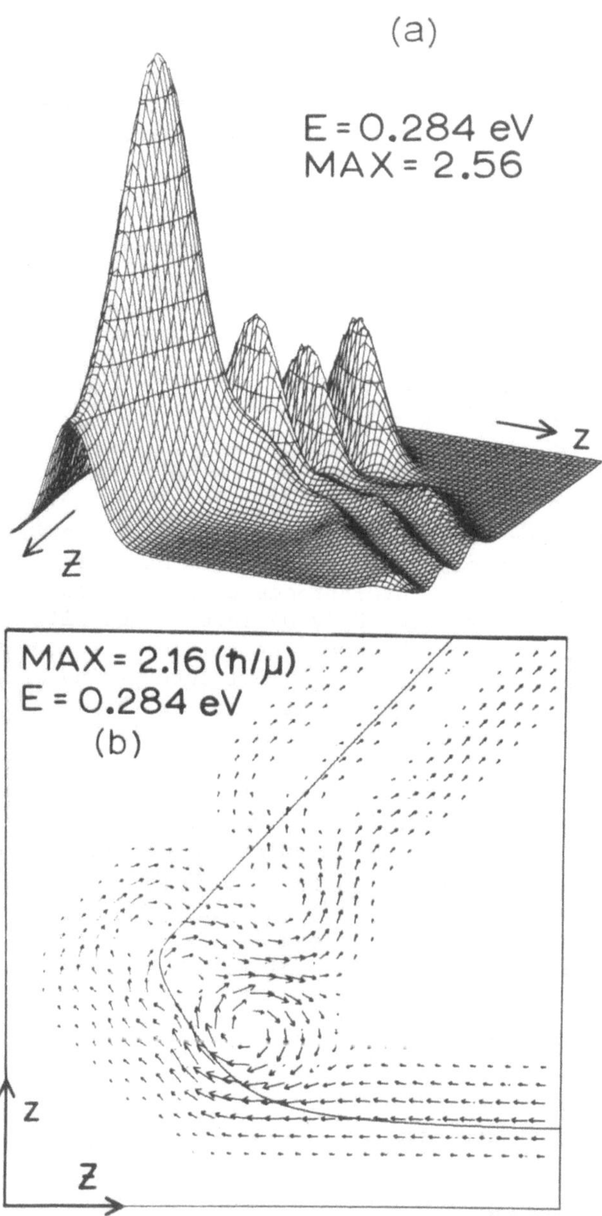

Figure 6. Transition state region (viewed from the F + H + H region) for the collinear F + H_2 reaction at 0.284 eV total energy: (a) probability density (the reactant channel is to the left, and the product channel emerges at the lower right); (b) probability flux, top view of the same region as part (a). The magnitude of the longest flux vector is 2.16 (h/μ).

with quantized whirlpools in the probability flux maps, and with buildup of $\Psi^*\Psi$ in the transition state region.

D. Resonance Models

Numerical solutions of the CC equations provide exact descriptions of resonance states in reactions, but little insight into the resonance mechanism is provided unless further analysis is performed. Several direct approaches to the influnece of resonance states on scattering processes are available,[28] of which one of the most widely quoted was derived by Feshbach. In attempting to remedy the lack of numerical applications of the Feshbach formalism to chemical reactions, we have found it useful[29] to study a number of "box models", one of which is shown in Fig. 7. There are two open and two asymptotically closed channels. The upper two channels temporarily open in the "transition state" region, $0 \leq R \leq L$. How does trapping of probability in the resonant states (associated with coupled channels 3 and 4) influence the reactive and inelastic open channel processes?

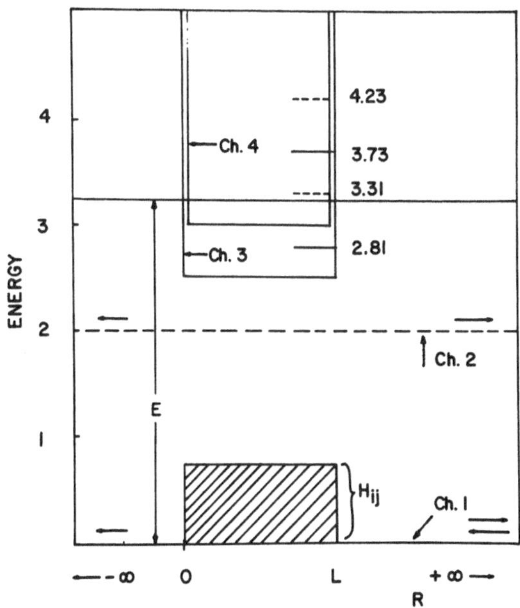

Figure 7. Channel potentials and coupling matrix elements for model resonance problems. Channels 1 and 2 are open at energy E. Input flux enters in channel 1. Channels 3 and 4 are asympotically closed, but temporarily open for $0 \leq R \leq L$. The lowest two energy levels in channels 3 and 4 are shown, for L = 4. H_{ij} is a typical coupling matrix element between channels i and j.

In matrix form, the close-coupled equations for the translational wavefunctions $(f_1(R),\cdots,f_N(R))$ are

$$\left[\underline{\underline{1}}(T_R-E)+\underline{\underline{\varepsilon}}+\underline{\underline{H}}\right]\underline{F}(R)=\underline{0} \tag{14}$$

where $\underline{\underline{\varepsilon}}$ (diagonal) contains the channel (vibrational) energies and $\underline{\underline{H}}(R)$ contains the channel coupling terms. This equation splits exactly into coupled equations for the open (channels 1,2) and closed (channels 3,4) components of $\underline{F}$:

$$\left[\underline{\underline{1}}_o(T_R-E)+\underline{\underline{\varepsilon}}_o+\underline{\underline{H}}_{oo}\right]\underline{F}_o(R)=-\underline{\underline{H}}_{oc}\underline{F}_c\,, \tag{15}$$

$$\left[\underline{\underline{1}}_c(T_R-E)+\underline{\underline{\varepsilon}}_c+\underline{\underline{H}}_{cc}\right]\underline{F}_c(R)=-\underline{\underline{H}}_{co}\underline{F}_o\,, \tag{16}$$

where the subscripts o or c denote open (dimension m=2) or closed (n=2) subspaces. Associated with the open channel problem are two associated equations for the homogeneous solution and for the Green function matrix:

$$\left[\underline{\underline{1}}_o(T_R-E)+\underline{\underline{\varepsilon}}_o+\underline{\underline{H}}_{oo}\right]\underline{F}_o^o=\underline{0}\,,$$

$$\left[\underline{\underline{1}}_o(T_R-E)+\underline{\underline{\varepsilon}}_o+\underline{\underline{H}}_{oo}\right]\underline{\underline{G}}_o(R,R')=\underline{\underline{\Delta}}(R-R'). \tag{17}$$

where $\underline{\underline{\Delta}}$ has a source function $\delta(R-R')$ in each diagonal element. In addition, the coupled closed channel problem has an associated bound-state problem with real eigenvalues $W_p(p=1,2\cdots)$,

$$\left[\underline{\underline{1}}_c(T_R-W_p)+\underline{\underline{\varepsilon}}_c+\underline{\underline{H}}_{cc}\right]\underline{\Phi}_p=\underline{0}, \tag{18}$$

in terms of which the set of square integrable functions $\underline{F}_c$ may be expanded,

$$\underline{F}_c(R)=\sum_{p=1}^{n}d_p(E)\underline{\Phi}_p(R). \tag{19}$$

An equation for $\underline{F}_o$ is easily derived from Eqs. (15) and (17) by noting that

$$\underline{F}_o(R)=\underline{F}_o^o(R)-\hat{\mathcal{G}}_o\underline{\underline{H}}_{oc}\underline{F}_c(R), \tag{20}$$

where $\hat{G}_o$ is the Green operator whose matrix kernal is $\underline{G}_o$ in Eq. (17). Inserting $\underline{F}_o$ into Eq. (16) then leads to the desired inhomogeneous equation for $\underline{F}_c$:

$$\left[\underline{\underline{1}}_c(T_R-E)+\underline{\underline{\varepsilon}}_c+\underline{\underline{H}}_{cc}-\underline{\underline{H}}_{co}\hat{\mathcal{G}}_o\underline{\underline{H}}_{oc}\right]\underline{F}_c=-\underline{\underline{H}}_{co}\underline{F}_o^o. \tag{21}$$

Now, substituting the expansion in Eq. (19) into this equation yields a system of linear inhomogeneous equations for the d_p's. With these coefficients known, the open channel solution vector is

$$\underline{F}_o(R) = \underline{F}_o^o(R) - \sum_p d_p(E) \int_{-\infty}^{+\infty} \underline{\underline{G}}_o(R,R') \underline{\underline{H}}_{oc}(R') \underline{\phi}_p(R') dR' \tag{22}$$

Finally, taking the limits of $\underline{F}_o$ as $R \to -\infty$ ("reactive scattering") or $R \to +\infty$ (inelastic or elastic scattering) leads to expressions for the S-matrix elements (coefficients of the outgoing waves):

$$\underline{S}_r = \underline{S}_r^o - \sum_p d_p(E) \int_{-\infty}^{+\infty} \underline{\underline{\eta}}_r \underline{\underline{H}}_{oc} \underline{\phi}_p dR' \tag{23}$$

$$\underline{S}_{nr} = \underline{S}_{nr}^o - \sum_p d_p(E) \int_{-\infty}^{+\infty} \underline{\underline{\eta}}_{nr} \underline{\underline{H}}_{oc} \underline{\phi}_p dR'$$

where $\underline{\underline{\eta}}_r$ and $\underline{\underline{\eta}}_{nr}$ are defined by taking the $R \to -\infty$ (reactive) or $R \to +\infty$ (nonreactive) limits of $\underline{\underline{G}}_o$, for example $\underline{\underline{G}}_o$ $(R \to -\infty, R') \to \underline{\Lambda}_r(R)\underline{\eta}_r(R')$, where $\underline{\Lambda}_r$ is diagonal, but $\underline{\underline{\eta}}_r$ is not. In Eq. (23), the first terms are contributions from "pure" open channel scattering (direct reaction mechanism), while the summations put in the effect of temporary excitation into the resonant states.

This procedure has been implemented for a number of model reactive resonance problems,[29] including the one shown in Fig. 7. The channel coupling matrices for $0 \leq R \leq L$ are

$$\underline{\underline{H}}_{oo} = \begin{bmatrix} 0.00 & 0.05 \\ 0.05 & 0.00 \end{bmatrix}, \quad \underline{\underline{H}}_{cc} = \begin{bmatrix} 0.00 & 0.30 \\ 0.30 & 0.00 \end{bmatrix},$$

$$\underline{\underline{H}}_{oc} = \underline{\underline{H}}_{co} = \begin{bmatrix} 0.40 & 0.40 \\ 0.40 & 0.40 \end{bmatrix}. \tag{24}$$

A basis set of five bound states was used in each of the two closed channels in order to variationally solve Eq. (19). For the box length L=4, the four lowest eigenvalues W_p are 2.668, 3.449, 3.593 and 4.374 in units where $\hbar = 1$ and $m = 1$. The actual resonance energies are closely predicted by the complex eigenvalues of the non-Hermitian matrix on the left side of Eq. (21). The four lowest eigenvalues are: (2.727, -0.001), (3.663, -0.012), (4.7369, -0.0619), (5.2195, -0.0268), where the imaginary part (second number) measures the resonance width.

The energy dependence of the four probabilities associated with the box model having two open nonreactive and two open reactive channels is shown in Fig. 8. The sharp resonance near E = 2.72 is due to density buildup in the resonant state built primarily from the lowest bound state in channel 3. The wider resonance near E = 3.66 arises from a resonant state built from the ground state in channel 4 and the first excited state in channel 3. It also appears from Fig. 8 that the resonance shapes are not well fit by the Breit-Wigner (BW) formula, which applies

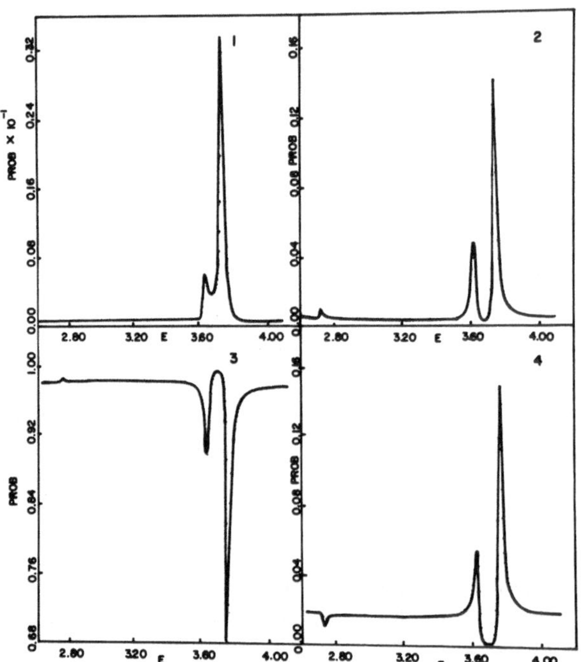

Figure 8. Transition probabilities between channel 1 (elastic channel) and all output channels (1 = ch. 1-right, 2 = ch. 2-right, 3 = ch. 1-left, 4 = ch. 2-left). The box model in Fig. 7 was employed, with L=4.

to isolated resonances. Deviations from the BW lineshape are due to interference effects between an isolated resonant process and the pure open channel processes or between different resonant processes. These different types of terms arise when the absolute squares of the S-matrix elements are computed from Eq. (23).

Resonance models of this type should help us to understand many aspects of resonance phenomena in reactive collisions, including the following:

a. formation of collective resonances due to the interaction of close bound states in different closed channels,

b. development of useful models for resonance shifts and widths, and their dependence upon the various H_{ij} coupling matrix elements,

c. the approach to the statistical limit as the resonance widths exceed their average spacings.

These and other aspects of interfering resonances in reactive collision models will be presented elsewhere.[29] Several other studies of resonance phenomena in molecular collisions have also appeared.[30]

III. ELECTRONIC NONADIABATIC COLLISIONS

A. Close-Coupling Theory

As might be expected, quantum mechanical studies of electornic nonadiabatic collision processes are much less numerous than for electronic adiabatic processes.[31] However, some progress has been made, as evidenced by computational results on the collinear $X(^2P$; halogen) + H_2, $H^+ + H_2$, and Ba + N_2O reactions,[32] by computational results for the three-dimensional $F(^2P) + H_2(D_2)$ inelastic collisions,[33-36] and by several theoretical formulations of the electronic-nuclear energy transfer problem.[37-39] Our interest here is in the quenching, by H_2, of electronically excited fluorine (the $^2P_{1/2}$ - $^2P_{3/2}$ splitting is 0.05 eV):

$$F(^2P_{1/2}) + H_2(j=0) \begin{cases} \nearrow F(^2P_{1/2}) + H_2(j'=0), \text{ elastic} \\ \rightarrow F(^2P_{3/2}) + H_2/j'=0,2), \text{ inelastic quenching} \\ \searrow FH('\Sigma;v'j') + H, \text{ reactive quenching .} \end{cases}$$

The reactive process has been modeled quantum mechanically,[34,35] and semiclassically[31,22] and the nonreactive processes have been treated within the close-coupling formalism.

The total Hamiltonian for this system partitions into body-fixed electronic and nuclear motion components, $H = \hat{H}_{el}$ (spin-free) + $\lambda L_F \cdot S_F + \hat{T}_{rot} + \hat{T}_{tr}$, where the second term is the spin-orbit interaction on F (with λ equal to the value for an isolated atom), and where the rotational Hamiltonian includes both molecular rotation (with angular momentum operator $\hat{j}$) and atom-molecule orbital rotation (with angular operator $\hat{\ell} = \hat{J} - \hat{J}_{el} - \hat{j}$). Vibrational motion is not explicitly included, since we are primarily interested in electronic-rotational energy transfer in the F - H_2 arrangement channel.

In order to provide a representation for $\hat{H}_{el}$, both a Hartree-Fock electronic basis, with added long range quadrupole interaction terms has been employed[33], and DIM basis functions[40] have been used by two other groups.[34-36,38,39]

In the DIM formulation of the electronic problem,[39] 24 polyatomic basis functions (built from six atomic ^2P F-atom states and two ^2S states on each H-atom), may be combined according to the angular momentum coupling scheme $(L_FS_F)J_F(S_AS_B)S_H(J_FS_H)J_eK_e \equiv \sigma K_e$ (where $(J_1J_2)J_3$ means that $\vec{J}_1 + \vec{J}_2 = \vec{J}_3$), where K_e is the electronic angular momentum along the BF z-axis (asymptotically, the F-H_2 vector). As a result, the 24 x 24 electronic matrix, with elements $\langle\sigma K_e|\hat{H}_{el}|\sigma' K_e\rangle$, may be generated at all values of the internuclear distances from six input diatomic potential curves. This electronic matrix is diagonal at large $R(F-H_2)$, since the coupling scheme is chosen to diagonilize $\hat{L}_F \cdot \hat{S}_F$.

In order to perform scattering calculations, the DIM electronic basis $|\sigma K_e\rangle$ must be augmented with nuclear rotational basis functions,[39] which in a space-fixed frame are the spherical harmonic products $|jm_j\rangle|\ell m_\ell\rangle$. The products of these three types of functions are then best combined to form eigenkets of the total angular momentum $\hat{J}^2$, where $\hat{J} = J_e + \hat{j} + \ell$. These "electrotational" functions are denoted $\Omega_\alpha^{JM}(\hat{\theta}\phi\chi\gamma e;s)$, where as in Sec. II, $\{\theta\phi\chi\}$ orients the nuclear triangle, γ bends the system away from collinear geometries, e represents the set of all body-fixed electronic coordinates, and where the diatomic electronic potential curves used to define $|\sigma K_e\rangle$ are evaluated at position s along the reaction coordinate. In addition, $\alpha = (j\ell)L_n\sigma(L_nJ_e)J$ denotes the nuclear-plus-electronic angular momentum coupling scheme. The index α is not a good quantum number at small $F-H_2$ distances. Unlike the situation with nuclear dynamics on Born-Oppenheimer adiabatic electronic surfaces, the total nuclear angular momentum, L_n, is not conserved.

At each value of s, the total scattering wavefunction for elastic channel α may be expanded in the set of electrotational basis functions,

$$\Psi_\alpha^{JM}(\theta\phi\chi\gamma e;s) = I\sum_{\alpha'=1}^{N} T_{\alpha'\alpha}^{J}(s)\,\Omega_{\alpha'}^{JM}(\theta\phi\chi\gamma e;s) \tag{25}$$

in which I is a scale factor. The CC equations then take the form

$$\left(\frac{d^2}{ds^2} - \frac{2\mu E}{\hbar^2}\right)T_{\alpha'\alpha}^{J}(s) = -\frac{2\mu}{\hbar^2}\sum_{\alpha''}\langle JM\alpha'|\hat{H}_{el}+\hat{T}_{rot}|JM\alpha''\rangle T_{\alpha''\alpha}^{J}(s) \tag{26}$$

The electronic Hamiltonian links the electrotational functions through the electronic matrix elements $\langle JM\alpha|\hat{H}_{el}|JM\alpha'\rangle$. It is interesting to note that even-j and odd-j molecular rotational states are allowed to mix, if they are associated with different electronic states[35] ($\sigma' \neq \sigma$). This will be seen in the next section to have a dramatic effect upon the adiabatic electronic-rotational energy correlation diagram at small $F-H_2$ distances.

B. $F(^2P) + H_2$ Quenching

The set of close-coupling equations have been numerically integrated[34-35] with the R-matrix propogation method.[14] At an intermediate stage of the integration, the adiabatic electrotational energies at each s are obtained by diagonalizing the coupling matrix in Eq. (26). The adiabatic energy correlation diagram for $J = \frac{1}{2}$ is shown in Fig. 9, for an electrotational basis containing j=0,2,4 H_2 states, with $^3\Sigma_u$ H_2 electronic states excluded. When the latter are included in the basis, the lowest curve looses its repulsive character in the region near s=0, and dips down toward the FH + H product valley (the reaction is exoergic by 1.37 eV).

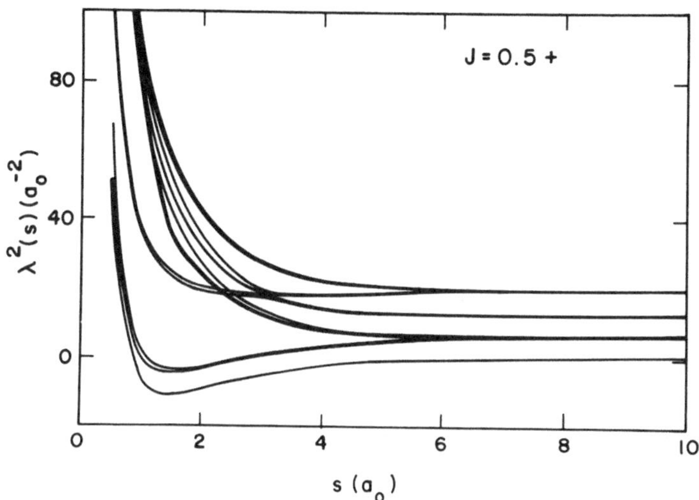

Figure 9. $F + H_2$ electronic-rotational energy correlation diagram for J = 1/2. The eigenvalues of the coupling matrix are plotted as a function of the reaction coordinate $|s|$. Starting at E=0 in the asymptotic region, the curves correspond to

$F(^2P_{3/2}) + H_2(j=0)$
$F(^2P_{1/2}) + H_2(j=0)$ } nearly degenerate
$F(^2P_{3/2}) + H_2(j=2)$
$F(^2P_{1/2}) + H_2(j=2)$
$F(^2P_{3/2}) + H_2(j=4)$

In this figure, the energies (λ^2) are in a_o^{-2} units, and are obtained by multiplying the energy (eV) by $2\mu/h^2 = 0.00779$.

The dependence upon J of the transition probabilities for the near-resonant quenching process (leading to $F(^2P_{3/2}) + H_2(j'=2)$) and the non-resonant process (leveling to j'=0) are shown in Fig. 10a, for E_{tr} = 0.02 eV. The interweaving of the probabilities for even and odd parities ($j+\ell$ = even or odd) is due to in-out alteration of avoided crossings in the adiabatic electrotational correltaion diagrams.[34,36] The energy dependence of the transition probabilities for J = 1/2 are shown in Fig. 10b, for electronic basis sets with and without $^3\Sigma_u^+$ H_2 electronic components. In addition, the reaction probabilities shown in this figure were estimated by applying "detailed quantum transition state theory",[41] for both initial spin-orbit states of fluorine.[35] The probability for the reactive quenching process is much smaller than for the inelastic quenching processes but rapidly increases as the reactant translational

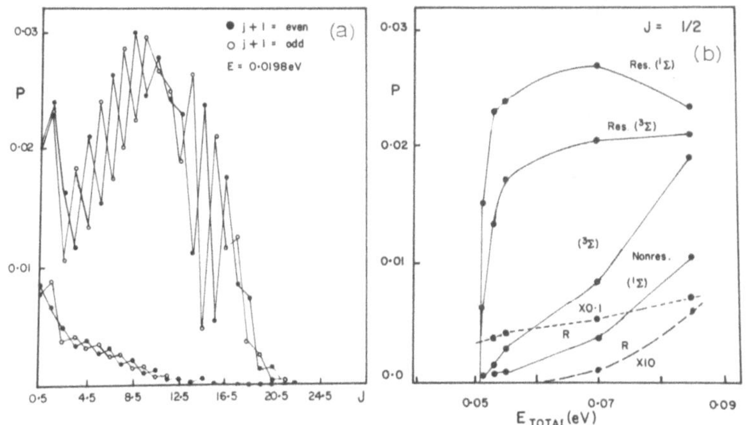

Figure 10. (a) Electronic quenching transition probabilities, $F(^2P_{1/2}) + H_2(j=0) \rightarrow F(^2P_{3/2}) + H_2(j')$, for $E_{trans} = 0.0198$ eV. The upper pair of curves are for near-resonant transitions (j'=2), the lower pair for non-resonant transitions (j'=0). Note the interweaving pattern for even and odd molecular parities ($j+\ell$ = even or odd, respectively).

(b) Transition probabilities when J = 1/2 for near-resonant (Res.) and nonresonant (Nonres.) quenching. Curves marked $^1\Sigma$ are for DIM bases including only $^1\Sigma_g^+$ H_2 states, while those marked $^3\Sigma$ include both $^1\Sigma_g^+$ and $^3\Sigma_u^-$ electronic components. The upper curve marked R is the reaction probability from $F(^2P_{3/2}) + H_2(j=0)$. The lower R curve is the reaction probability from $F(^2P_{1/2}) + H_2(j=0)$. Values on the upper R curve have been multiplied by 0.1 for plotting; the lower R values have been scaled up by a factor of 10 for plotting.

energy increases. In addition, the figures clearly show the dominance of the near-resonant inelastic process (leading to $H_2(j'=2)$) over the non-resonant one. The ratio of total cross sections for these inelastic processes ($\sigma_{near-res}/\sigma_{non-res}$) varies from 40 to 100 as E_{tr} increases over this energy range. Experimental studies to test and extend these quantum studies would be most welcome. In addition, model quantum mechanical studies (e.g. the distorted wave Born approximation[42]) of electronic energy transfer should be useful.

IV. CONCLUSIONS

Quantum mechanical theory and results have been presented for three-dimensional A + BC type chemical reactions. During the past several years, new results (reaction probabilities, cross sections, and transition state dynamic details) have been obtained on the $H + H_2$ and $F + H_2$ electronic adiabatic reactions, and an electronic-rotational energy transfer for $F + H_2$ reactant geometries. However, the lack of excited state potential surfaces, and intersurface coupling matrix elements will hinder the application of these methods to other reactive electronic energy transfer processes. In addition, channel reduction schemes, of which J_z-conserving is an example, need to be explored in more detail, since large channel size is one of the major roadblocks which must be overcome (possibly through model effective Hamiltonians to simulate the effects of the "neglected" channels) before accurate extension of the close-coupling formalism can be made to other reactions. In spite of these problems, it seems very likely that formalisms of the general type discussed here will be extended over the next decade to three-dimensional reactive processes of the following type: (1) one path reactions ($Li + HF \rightarrow LiF + H$), (2) reactions with bent intermediates, (3) asymmetric two-path reactions (F +HD), (4) reactions in laser fields,[43] (5) dissociative collisions, (6) diatom-diatom reactions.

For many discussions on reactive collisions over the past several years, I wish to thank my collaborators: Allan Elkowitz, Susan Harms, Robert Walker, Michael Redmon, Joe McNutt and Curtis Shoemaker.

REFERENCES

1. Elkowitz, A. B. and Wyatt, R. E.: 1975, J. Chem. Phys. 62, p. 2504; 63, p. 702.
2. Schatz, G. C. and Kuppermann, A.: 1975, J. Chem. Phys. 62, p. 2502; 1976, 65, p. 4668.
3. Walker, R. B., Stechel, E. B. and Light, J. C.: 1978, J. Chem. Phys. 69, p. 2922.
4. Light, J. C., in Bernstein, R. B.(ed.), "Atom-Molecule Collision Theory: A Guide for the Experimentalist," (Plenum, New York, 1979) p. 467.
5. Micha, D. A.: 1975, Adv. Chem. Phys 31, p. 7.
6. Truhlar, D. G. and Wyatt, R. E.: 1976, Ann. Rev. Phys. Chem. 27, p. 1.
7. Wyatt, R. E., in Bernstein, R. B., op. cit., p. 567.
8. Wyatt, R. E., in Bernstein, R. B., op. cit., p. 477; and Wyatt, R. E., in Brooks, P. R. and Hayes, E. F. (eds.) "ACS Symp. Series" 1977, 56, p. 185.
9. Redmon, M. J. and Wyatt, R. E.: 1975, Int. J. Quantum Chem. Symp. 9, p. 403; 1977, Int. J. Quantum Chem. Symp. 11, p. 343.
10. Redmon, M. J. and Wyatt, R. E.: 1979, Chem. Phys. Lett. 63, p. 209.
11. Marcus, R. A.: 1968, J. Chem. Phys. 49, p. 2610.

12. Harms, S. H. and Wyatt, R. E.: 1975, J. Chem. Phys. 62, p. 3173.
13. Harms, S. H., Elkowitz, A. B. and Wyatt, R. E.: 1976, Mol. Phys. 31, p. 177.
14. Light, J. C. and Walker, R. B.: 1976, J. Chem. Phys 65, p. 4272.
15. Kouri, D. J.: Bernstein, R. B. (ed.), op. cit., p. 302.
16. Elkowitz, A. B. and Wyatt, R. E.: 1976, Mol. Phys. 31, p. 189.
17. Wu, S. F., Johnson, B. R. and Levine, R. D.: 1973, Mol. Phys. 25, p. 839.
18. Schatz, G. C., Bowman, J. M. and Kuppermann, A.: 1973, J. Chem. Phys. 38, p. 4023.
19. Schatz, G. C., Bowman, and Kuppermann, A.: 1975, J. Chem. Phys. 63, p. 674; 1975, J. Chem. Phys. 63, p. 685.
20. Connor, J. N. L., Jakubetz, W. and Manz, J.: 1975, Mol. Phys. 29, p. 347.
21. The form of the M5 potential is in Muckerman, J. T.: 1972, J. Chem. Phys. 56, p. 2997. Parameters are from Ref. 19.
22. Komornicke, A., Morokuma, K. and George, T. F.: 1977, J. Chem. Phys. 67, p. 5012.
23. Lee, Y. T.: private communication.
24. Schafer, T. P., Siska, P. E., Parson, J. M., Tully, F. P., Wong, Y. C. and Lee, Y. T.: 1970, J. Chem. Phys. 53, p. 3385.
25. Hutchinson, J. S. and Wyatt, R. E.: 1979, J. Chem. Phys. 70, p. 3509.
26. Latham, S. L., McNutt, J. F., Wyatt, R. E. and Redmon, M. J.: 1978, J. Chem. Phys. 69, p. 3746.
27. Hirscheelder, J. O., Goebel, C. J. and Bruch, L. W.: 1974, J. Chem. Phys 61, p. 5456.
28. Feshbach, H.: 1958, Ann. Phys. 5, p. 357; 1962, Ann. Phys. 19, p. 287. Also see: Yamabe, T., Tachibana, A. and Kudui, K.: 1978, Adv. Quantum Chem. 11, p. 195.
29. Shoemaker, C. and Wyatt, R. E.: to be published.
30. See the following, and references therein: Levine
Levine, R. D. and Wu, S. F.: 1971, Chem. Phys. Lett. 11, p. 557.
Schatz, G. C. and Kuppermann, A.: 1973, J. Chem. Phys. 59, p. 964.
Liedtke, R. C., Knirk, D. L. and Hayes, E. F.: 1977, Int. J. Quantum Chem. Symp. 11, p. 337.
31. Tully, J. C.: Miller, W. H., (ed.), "Modern Theoretical Chemistry: Molecular Collisions, Part B", (Plenum, New York, 1970); and Tully, J. C. in Brooks, P. R. and Hayes, E. F.: 1977, ACS Symp. Ser. 56, p. 206.
32. See the following, and references therein:
Top, Z. H. and Baer, M.: 1977, Chem. Phys. 25, p. 1.
Top, A. H. and Baer, M.: 1976, Chem. Phys. Lett. 39, p. 134.
Zimmerman, J. H. and George, T. F.: 1975, J. Chem. Phys. 63, p. 2109.
DeVries, P. L. and George, T. F.: 1977, J. Chem. Phys. 66, p. 2421.
Bowman, M., Leasure, S.,and Kuppermann, A.: 1976, Chem. Phys. Lett. 43, p. 374.
33. Rebentrost, F. and Lester, W. A. Jr.: 1976, J. Chem. Phys. 64, p. 4223.

34. Wyatt, R. E. and Walker, R. B.: 1979, J. Chem. Phys. 70, p. 1501.
35. Wyatt, R. E.: 1979, Chem. PHys. Lett. 63, p. 503.
36. DeVries, P. L. and George, T. F.: 1979, Chem. Phys. Lett. 63, p. 240.
37. Baer, M.: 1976, Chem. Phys. 15, p. 49.
38. DeVries, P. L. and George, T. F.: 1977, J. Chem. Phys. 67, p. 1293.
39. Miller, D. L. and Wyatt, R. E.: 1977, J. Chem. Phys. 67, p. 1302.
40. Ellison, F. O.: 1963, J. Am. Chem. Soc. 85, p. 3540; Ellison, F. O. and Patel, J. C.: 1964, ibid. 86, p. 2115.
41. Light, J. C. and Altenberger-Siczek, A.: 1975, Chem. Phys. Letts. 30, p. 195.
42. McNutt, J. F. and Wyatt, R. E.: to be published.
43. See the following, and references therein:
DeVries, P. L. and George, T. F.: 1979, J. Chem. Phys. 71, p. 1543.
Light, J. C. and Altenberger-Siczek, A.: 1979, J. Chem. Phys. 70, p. 4108.

34. Wy[illegible] [illegible] and [illegible]
35. Wy[illegible]
36. De[illegible]

SYMPOSIUM III. MOLECULAR INTERACTIONS

Chairman : J. Koutecky

Institut für Physikalische Chemie der
Freien Universität Berlin
Berlin, Germany

INTRODUCTORY REMARKS TO SYMPOSIUM 3: MOLECULAR INTERACTIONS

Jaroslav Koutecký
Institut fuer Physikalische Chemie, Freie Universität Berlin, 1 Berlin 33, West Germany

Due to the fact that quantum chemistry is just quantum physics applied to the field of chemistry and because chemistry is concerned primarily with interactions among molecules, the study of isolated molecules should be taken as a valuable preparation for predictions concerning molecular interactions. In order to appropriately use the results of quantum chemistry on isolated molecules, it is of great importance to investigate some features of molecular interactions. Therefore, the Symposium 3 of the Third International Congress of Quantum Chemistry is devoted to this topic. The three lectures in this Symposium can be taken as three bridges between the quantum chemistry and direct considerations on molecular interactions.

The first bridge connecting the theory with the experimental fact is represented by the report of Prof. Y.T. Lee, who will describe the experiments on reaction dynamics using crossed molecular beams method. This method makes it possible to investigate chemical reactions between molecules in well defined molecular states and to determine the molecular states of products in very exact way as well. As an example the simple reactions $F_2 + H_2$ and $F_2 + D_2$ have been chosen.

Prof. R. Marcus will discuss intramolecular dynamics paying special attention to some similarities with classical physics. Analogical to their classical physics counterpart, there exist states of vibrational motion in molecules which can be considered as stochastic ones if the vibration energies are high enough. Prof. Marcus' talk is therefore a bridge between the concepts of classical physics and the concepts of quantum chemistry.

In the last contribution of this Symposium Prof. P. Craig will report on the interactions between chiral molecules.

K. Fukui and B. Pullman (eds.), Horizons of Quantum Chemistry, 89–90.

The intention is to show that the well known concepts of chirality and chiral discrimination are also useful for the investigations concerning molecular interactions. This contribution can be considered as a bridge between quantum chemistry and profound geometrical considerations which for example are, as it is well known, of crucial importance in biochemical interactions.

The lectures of this Symposium show how it is fruitful to be aware of important connections and contacts between the quantum chemistry of isolated molecules and other fields of molecular science and theoretical chemistry.

MOLECULAR BEAM STUDIES OF REACTION DYNAMICS OF F + H_2, D_2

R.K. Sparks, C.C. Hayden, K. Shobatake
D.M. Neumark, and Y.T. Lee
Materials and Molecular Research Division
Lawrence Berkeley Laboratory and
Department of Chemistry
University of California
Berkeley, California 94720 USA

ABSTRACT

Reactions of F + H_2, D_2 have attracted extensive theoretical and experimental attention in recent years and will undoubtedly become one of the prototype chemical reactions in the future for the detailed understanding of reaction dynamics through meaningful comparison between theoretical calculations and experimental observations.

In this paper, recent state resolved high resolution crossed molecular beams studies of these reactions are discussed. The experimental results reveal such detailed information as the dependence of the angular distribution of products on initial collision energies and vibrational quantum states of products and give indication of recently predicted quantum mechanical resonance phenomena.

INTRODUCTION

Recent advances in quantum chemistry have made it possible to understand some important features of the reaction dynamics of such simple systems as $H + H_2 \rightarrow H_2 + H$ entirely based on quantum mechanical calculations.[1] The extension of this remarkable achievement to other simple hydrogen containing triatomic systems, especially the calculation of accurate ab initio potential energy surfaces, is progressing rapidly, although full three dimensional quantal calculations still await the development of more efficient methods. For the understanding of reaction dynamics of hydrogen containing systems, quantal scattering calculations are very important, since the significant quantum effects of tunneling and resonance phenomena which appear in these systems preclude the usage of classical trajectory calculations as a means of obtaining reliable information on the reaction dynamics. In the development of efficient approximation methods for carrying out three dimensional quantal calculations, comparison with accurate experimental results is necessary when the results of exact theoretical calculations are not available.

K. Fukui and B. Pullman (eds.), Horizons of Quantum Chemistry, 91–105.

One of the hydrogen-containing three atom systems which has already attracted extensive experimental and theoretical attention in the past and will undoubtedly become the prototype reaction in the detailed understanding of reaction dynamics is the reaction of fluorine atoms with hydrogen molecules. The calculation of a limited ab initio potential energy surface of chemical accuracy has been carried out.[2] The basic features of this calculation were adopted in the determination of more complete semiempirical potential energy surfaces which have been utilized in extensive classical trajectory calculations.[3,4] In exact one dimensional scattering calculations,[5,6] interesting quantum resonance phenomena were predicted at certain specific collision energies while a very recent three dimensional calculation using an angular momentum decoupling approximation[7] has provided some important information regarding resonance phenomena in three dimensional reactive scattering. The advances in recent years of various experimental approaches, such as chemical lasers,[8] chemiluminescence methods[9] and crossed molecular beams experiments,[10] have not only revealed detailed information on product state distributions, but also have provided angular distributions of products in various quantum states at well defined collision energies. These results and expected improvements in the near future will certainly provide an unprecedented opportunity for making very meaningful comparisons between theoretical calculations and experimental observations.

In particular, the experimental observation of resonance phenomena via their effect on the state and angular distributions of the products under well-defined initial conditions could provide a very stringent test of the accuracy of theoretical calculations. Since the resonance phenomena in chemical reactions are very sensitive to the potential energy hypersurface at the critical configuration of the reaction intermediate, the observation of resonance phenomena at specific collision energies in the laboratory provides an opportunity to evaluate the reliability of the potential energy surface in the close contact region.

In one dimensional quantum mechanical calculations, a sharp resonance has been predicted for the production of HF(v=2) at a collision energy of ~1 kcal/mole.[5,6] This sharp enhancement of the production of HF(v=2) at a specific collision energy was not observed in recent three dimensional calculations,[7] although a similar resonance phenomenon in the production of HF(v=2) predicted in the one dimensional calculation is still evident as collision energies are varied at a fixed orbital angular momentum of reactants. Because of both the shift of the resonance to higher collision energies when the orbital angular momentum is increased and the averaging of the distribution of orbital angular momenta in three dimensional reactive scattering, no pronounced variation of the state distribution as a function of collision energies is predicted in spite of the presence of resonance phenomena. This means that the measurement of state distributions as a function of collision energies will not be the appropriate method for the search for resonance phenomena. On the other hand, in the energy range where the resonance is occurring, there is a substantial variation in the reaction probability as a function of orbital angular momentum, or, equivalently, of impact parameter, for collisions at a given initial relative velocity. Since the angular distribution of product molecules to a large extent reflects the dependence of reaction probability on impact parameter, especially for a system with the lowest entrance barrier for the collinear approach, the search for unusual features in the angular distributions of various quantum states of HF measured in a high resolution cross molecular beams experiment is probably the best approach for finding reactive resonance phenomena in the laboratory.

In this paper, recent state resolved high resolution crossed molecular beams studies on F + H_2 → HF + H and F + D_2 → DF + D carried out at several collision energies will be discussed. The results reveal such detailed information as the dependence of the angular distribution of product molecules on both collision energy and product vibrational quantum state. These measurements illustrate the current state of the art of crossed molecular beams studies of reactive scattering and reflect predicted quantum mechanical resonances as well as other interesting features in the experimental observations of reactive scatterings.

EXPERIMENTAL

The experiment is performed by crossing two supersonic molecular beams, one of atomic fluorine seeded in N_2, and one of H_2 or D_2, in a newly designed high resolution universal crossed molecular beams machine. The two beams cross at 90° in a well-defined interaction region. The detector, a triply-differentially pumped electron-impact ionization mass spectrometer, is rotated about the interaction region in order to obtain the product angular distributions. The product velocity distributions are determined by measuring time-of-flight for molecules passing through a rapidly spinning slotted disc to the ionizer.

The design of the new apparatus closely follows that of the original universal crossed molecular beam machine described by Lee et al.,[11] but several modifications extend its capabilities beyond those of the original design. The detector is suspended from a 35" diameter rotating chamber which allows one to obtain a 30 cm flight path in contrast to the 17 cm flight path in the original 25" design. This nearly doubles the resolution of the time-of-flight spectra and consequently allows the product state distribution to be more accurately determined than was previously possible. In addition, the main chamber has been enlarged in order to increase the laboratory angular scan range from 115° to 170°.

Other modifications include a set of removable precision-machined apertures for the detector slits to allow the detector angular resolution to be varied simply by inserting a different set of apertures. Extreme care was taken in the choice and processing of all materials used in the three ultra-high vacuum detector regions, which has resulted in a background count rate at most masses that is nearly a factor of ten lower than in any of our previous machines. Also, a newly designed 1 μs per channel multi-channel scaler compatible with the CAMAC standard has been utilized for the time-of-flight analysis in conjunction with a new on-line computer system.

The fluorine atom beam was produced by thermal dissociation of a 3% mixture of F_2 in N_2 at 700 torr in a nickel oven heated by thermocoax[12] heating elements to a temperature of 706°C. The heated gases expand through a 0.15 mm nozzle at the tip of the oven. A 0.66 mm diameter skimmer was used with a nozzle-skimmer distance of 6 mm. The beam was collimated in a second chamber to an angular spread of 2.2°. The nozzle to interaction center distance was 5.5 cm. Nitrogen was used as a seed gas to avoid the increased background at mass 20 (HF) that would have resulted from the use of Ar or Ne. Using He as a seed gas resulted in such high F atom velocities that the orientation of the resulting Newton diagram was unfavorable for observing many important center-of-mass angles. The F atom beam obtained had a peak velocity of 13.7×10^4 cm and a FWHM velocity spread of approximately 25%.

Hydrogen and deuterium beams were produced by a very high pressure expansion of the gases through a 0.030 mm diameter nozzle. The nozzle-skimmer region was pumped by a Varian 10" NHS diffusion pump backed by a Leybold Heraeus WS-500 Roots blower package. The nozzle skimmer distance used was approximately 10 mm and was externally adjusted to allow maximization of beam intensity. The skimmer itself was constructed of electro-formed nickel[13] and has the advantages of very sharp edges and very small included angles. The beam was collimated to a width of 2.2° and the nozzle was located at a distance of 6.5 cm from the scattering center. The collision energy of the experiment was varied by changing the stagnation temperature of the D_2 or H_2 beam. The beam source was heated with a short section of thermocoax wire powered by a regulated D.C. power supply. At the higher temperatures, the stagnation pressures behind the nozzle were increased in order to maintain number density, and thus intensity and high quality ot expansion. The operating conditions of the H_2/D_2 beam source for the four different sets of data to be presented here are shown in Table I along with the most probable collision energy for each condition.

Product velocity analysis for the $F + D_2$ experiments was performed by single shot time-of-flight. The single shot time-of-flight wheel consisted of a 0.13 mm thick stainless steel disk of 17.5 cm diameter. Eight slots, each 1.5 mm wide, were machined on its perimeter. A reducing slit, also 1.5 mm wide, was mounted to the front of the detector. The wheel was rotated at a speed of 200 Hz. Data was collected for 600 μs following each wheel opening using a multichannel scalar channel width of 5 μs (120 channels). Typical counting times were two hours at each angle.

Table I. H_2/D_2 Operating Conditions

	Collision Energy kcal/mole	Temperature °C	Stagnation Pressure (Atm)	Peak Velocity cm/sec
H_2	2.00	32°C	27	27.2 x 10^4
	3.17	219°C	40	35.7 x 10^4
D_2	2.34	32°C	27	20.9 x 10^4
	4.5	421°C	54	39.9 x 10^4

Velocity distributions for the F + H_2 experiment were obtained by cross-correlation time-of-flight. Our particular implementation of this method has been described before.[14] Cross-correlation was necessary for F + H_2 due to the much higher background at mass 20 (HF) than at mass 21 (DF). However, it is not an appropriate technique for a system such as F + D_2, since, in that case, the signal to background ratio is sufficiently high such that cross-correlation would actually reduce the quality of the data.[15] Additionally, we are presently limited by wheel design and wheel speed to a channel width of 10 μs with cross-correlation. Data for F + H_2 was taken with this lower limit channel time of 10 μs corresponding to a wheel speed of 392 Hz with 255 channels on the wheel. Typical counting times were approximately 70 minutes at each angle.

Angular distributions were obtained by modulation of the H_2/D_2 beam with a 150 Hz tuning fork chopper. Data was collected for equal times during the open and closed portion of the tuning fork cycle in a dual channel NIM scalar. The difference of the counts in the two channels constitutes the angular scan signal. Typical counting times for the angular scans were 2 minutes at each angle. Typical signal near the maximum in the angular distribution for F + H_2 would be about 6000 counts with a background of 19,000 counts. For F + D_2 we usually achieved 1700 counts of signal with about 120 counts of background. These numbers represent about one sixth of the actual count rate due to our intentionally low duty cycle of the tuning fork chopper. Detector aperture slits were set to achieve a laboratory angular resolution of 1° for both angular scans and the TOF data.

RESULTS AND ANALYSIS

The measured laboratory angular distributions for F + H_2 at 2.00 and 3.17 kcal/mole collision energy are shown in Figs. 1 and 2 along with the calculated best fits. The nominal Newton diagrams for each system are also shown. The corresponding data and Newton diagrams for F + D_2 at collision energies of 2.34 and 4.51 kcal/mole are shown in Figs. 3 and 4.

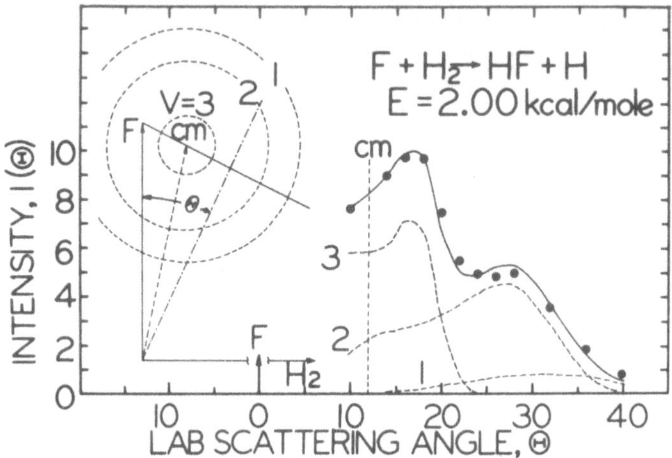

Fig. 1. Experimental LAB angular distribution of the HF product from the F + H_2 reaction at 2.00 kcal/mole.

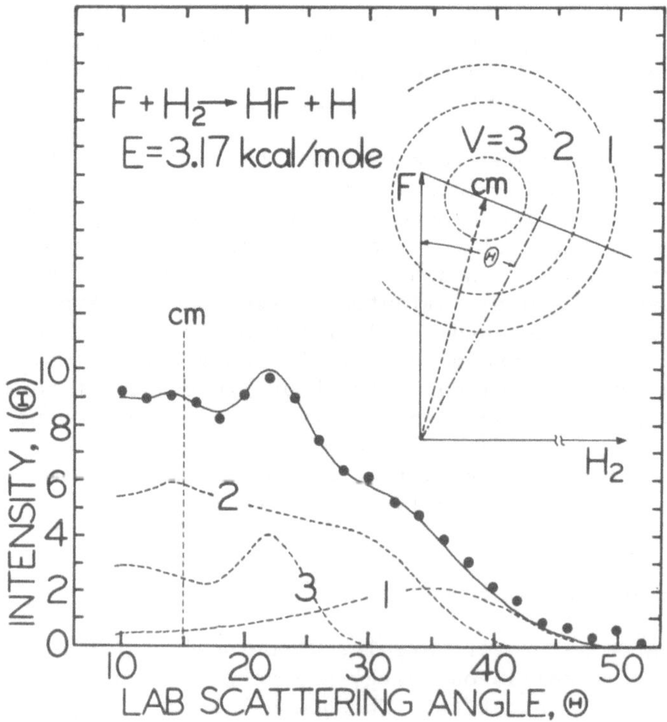

Fig. 2. Experimental LAB angular distribution of the HF product from the F + H_2 reaction at 3.17 kcal/mole.

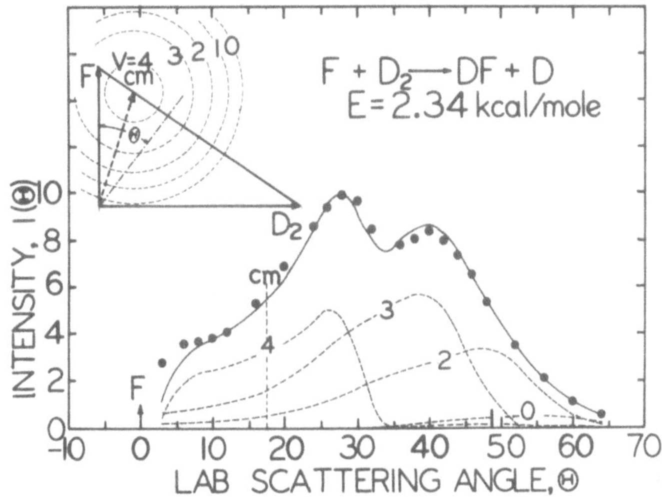

Fig. 3. Experimental LAB angular distribution of the DF product from the F + D_2 reaction at 2.34 kcal/mole.

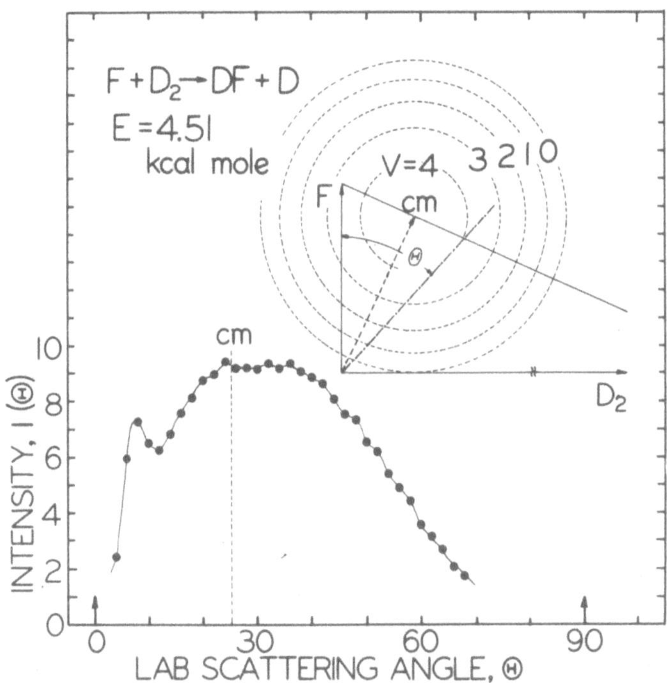

Fig. 4. Experimental LAB angular distribution of the DF prdouct from the F + D_2 reaction at 4.51 kcal/mole.

Time-of-flight data for all of these systems has been obtained. TOF spectra were generally taken for every other angle at which angular data was taken, however, over regions in which interesting effects were anticipated, spectra were taken for every angle at which angular scan data was taken. Angles with less than one fourth of the total intensity at the peak of the angular distribution were not considered profitable for TOF analysis. This precedure resulted in 11 to 18 TOF spectra for each system. Figure 5 shows the experimental TOF data for F + H_2 at 2.00 kcal/mole collision energy. The plots also show our calculated best fit to the data, including the separate contributions of the various vibrational states.

Data for both of the F + H_2 systems and for the lower energy F + D_2 have been analyzed to obtain the center-of-mass (C.M.) product distribution. We have used a forward convolution trial and error fitting technique in which the C.M. angular and energy distribution is input as a trial function. The corresponding laboratory angular and velocity distributions are then calculated from it and compared to the experimental data. The original trial function is adjusted and the process repeated until a satisfactory fit is obtained simultaneously to both the TOF spectra and the angular scan. The appropriate equations and basic methodology of forward convolution have been well discussed elsewhere.[15] We, however, found it necessary to write an entirely new computer program for the actual calculation. Our very large quantity of data forced us to use direct graphical output of our calculated distributions. We considered completely different angular and energy distributions for each vibrational state, as well as having the energy distribution of each state being angular dependent.

The C.M. angle-recoil energy distribution that we have used is a generalized RRK form with separate parameterization for each vibrational state. The functional form used is:

$$I_{C.M.}(E,\theta) = \sum_{v=0}^{n} C_v(\theta)N_v(\theta)\cdot\begin{cases}(E-E_{min}(\theta,v))^{\alpha(\theta,v)}(E_{max}(\theta,v)-E)^{\beta(v)}\\ 0 \text{ if } E < E_{min}(\theta,v) \text{ or } E > E_{max}(\theta,v)\end{cases} \quad (1)$$

where $C_v(\theta)$ is a relative angular intensity factor, which is input in point form every 10° and interpolated linearly by the program for points between 10° intervals. $N_v(\theta)$ is a normalization factor calculated by the program to normalize the rest of the function to an area of 1 at each angle. $\alpha(\theta,v)$ is input in point form and interpolated in the same manner as $C_v(\theta)$. E_{min} and E_{max} were defined as circles in (E,θ) space with a given radius in E and a given displacement along the 0°-180° axis. Usually the radius and displacement for E_{min} were chosen as 0, while the displacement for E_{max} was usually chosen as zero with the radius being set at the sum of the collision energy and the exothermicity minus the vibrational energy of the particular state. The

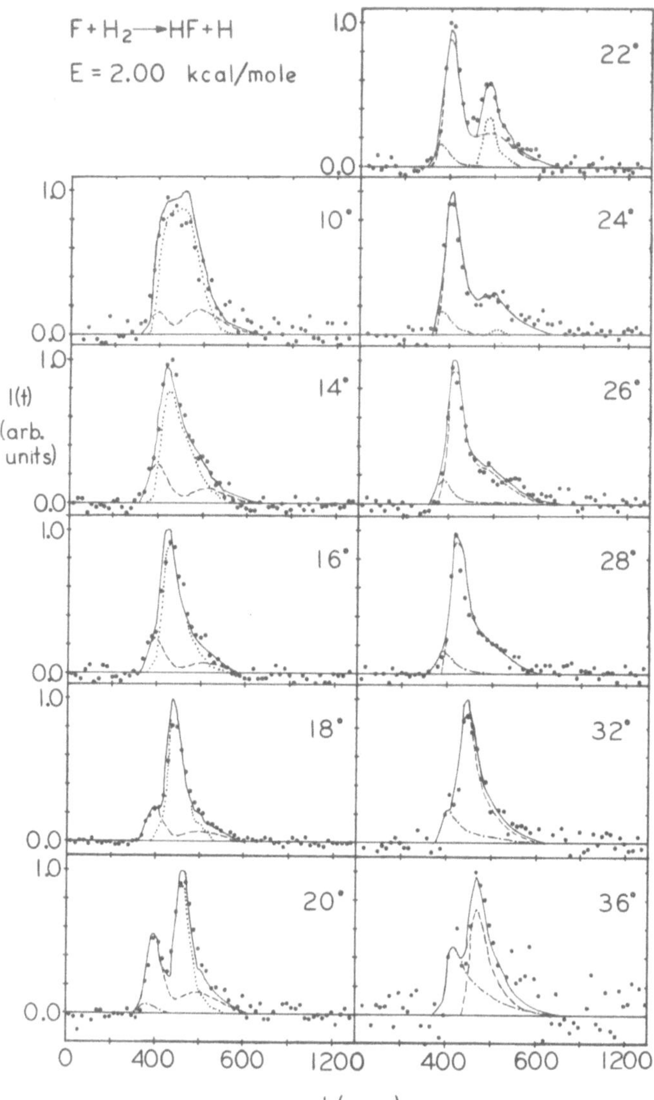

Fig. 5. Time-of-flight analysis of the HF product from F + H_2 reaction at 2.00 kcal/mole. Plots show experimental LAB TOF spectra at indicated angles along with the best fits calculated from center-of-mass angular and velocity distributions for various vibrational states of HF. (●) Experimental. (———) Calculated best fit summed over contributions from v=1,2, and 3. (• ——) v=1. (—— ——) v = 2. (•••••) v=3.

parameter β was fixed for each vibrational state. The total distribution, then, represents the entire product flux recoil energy distribution while the contribution from each vibrational state, in principle represents the flux as a function of rotational excitation of that vibrational state.

Thus the final result of our data fitting analysis is a center-of-mass flux distribution as a function of angle and product recoil energy. In order to represent this distribution in the usual form superimposed on a Newton diagram it must be converted from an energy space to a velocity space distribution. The transformation is straightforward and the results are shown in Figs. 6-8.

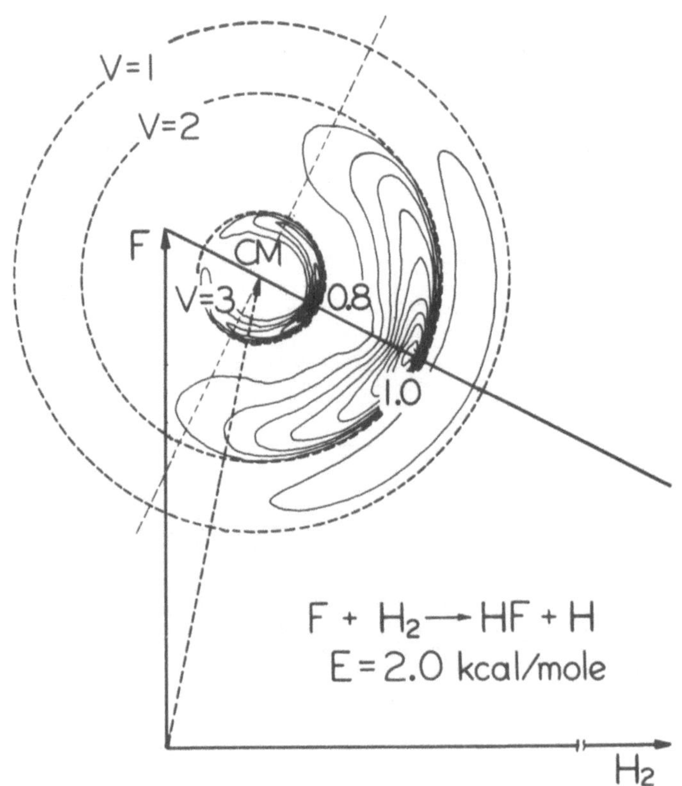

Fig. 6. Center-of-mass velocity-space contour plot of HF product distribution from F + H_2 reaction at 2.00 kcal/mole.

Figures 6 and 7 show the velocity space contour maps of our best fit distribution for F + H_2 at 2.00 and 3.17 kcal/mole respectively. Figure 8 shows the corresponding information for F + D_2 at 2.34

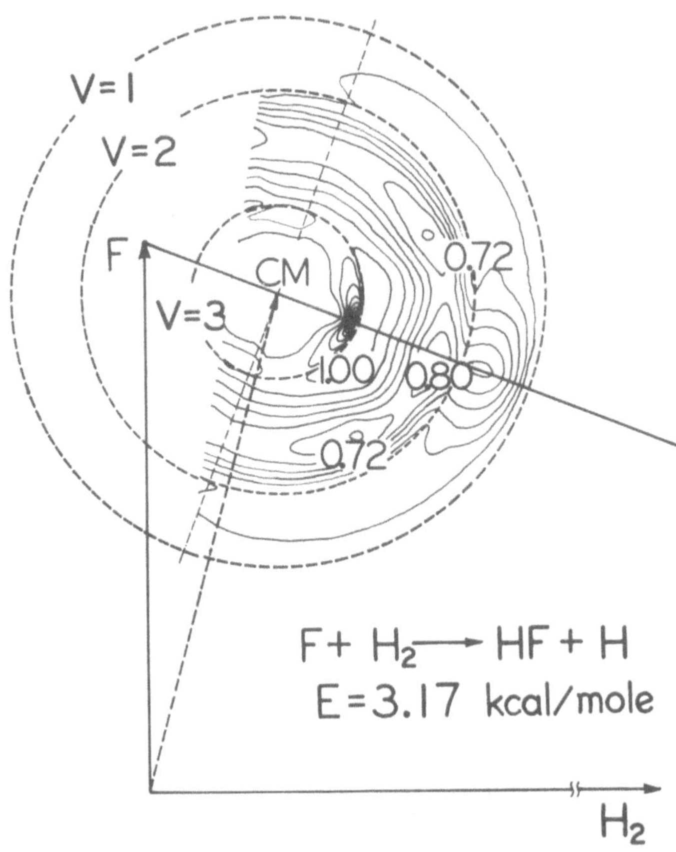

Fig. 7. Center-of-mass velocity space contour plot of HF product distribution from F + H_2 reaction at 3.17 kcal/mole.

kcal/mole. When transformed to laboratory angular and velocity space, these contour maps correspond to the total calculated distributions shown in the angular scan plots of Figs. 1-3 and the velocity distribution plots of Fig. 5.

DISCUSSION

The dynamics of the F + H_2 and F + D_2 reactions produce several features which lend themselves well to study by the crossed molecular beams method. The large product vibrational spacings in conjunction with narrow rotational energy distributions in many cases allow the vibrational states of the HF and DF products to be identified from distinct peaks which appear in the measurements of laboratory velocity. The potential energy barrier to reaction is lowest in the configuration

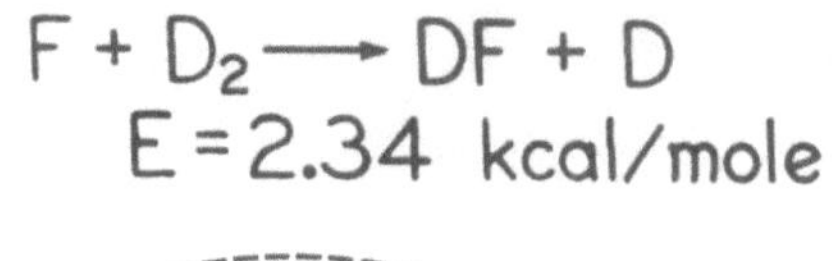

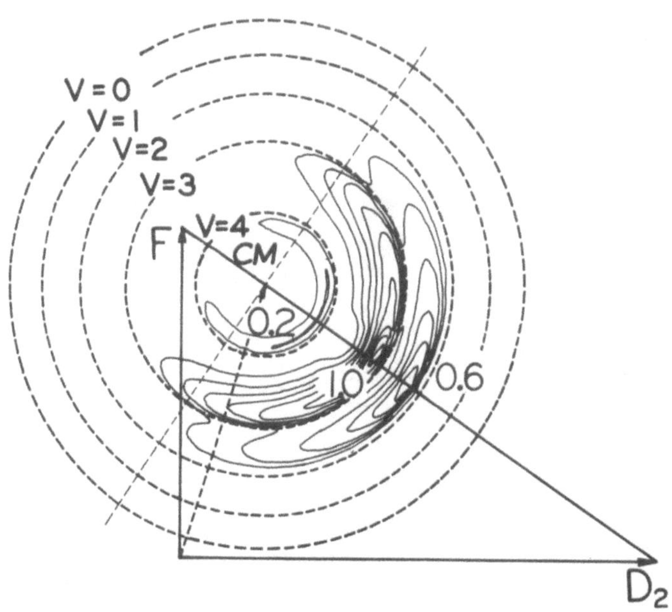

Fig. 8. Center-of-mass velocity-space contour plot of DF product distribution from F + D_2 reaction at 2.34 kcal/mole.

where the fluorine atom approaches collinear to the axis of the hydrogen molecule. At low collision energies this is expected to cause the products to scatter predominantly along the relative velocity vector of the reactants, and consequently peaks corresponding to individual vibrational states are observed in the laboratory angular distributions. Thus, both the laboratory velocity measurements and angular distributions provide detailed information on product state distributions and the two measurements effectively complement one another.

At low collision energies the angular distributions of HF and DF in the center of mass coordinate system shown in Fig. 6 and Fig. 8 are found to peak backward with respect to the direction of the incident fluorine atom. This is consistent with the nature of the potential energy surfaces derived from ab initio and semi-empirical calculations for F + H_2. The potential energy barrier for this reaction depends on the bending angle of H-H-F. The barrier is lowest for the collinear configuration, with a value of about 1 kcal/mole. Consequently, only those collisions involving a nearly collinear configuration of the two hydrogen atoms and the fluorine atom contribute to the formation of products, and these products are predominantly backward scattered.

The center-of-mass angular and velocity distribution for HF product from the $F + H_2 \rightarrow HF + H$ reaction at a collision energy of 3.17 kcal/mole is shown as a contour map in Fig. 7. At this collision energy the center-of-mass angular distribution differs markedly from that at the lower energy. The distributions for different vibrational states seem to be entirely decoupled. The products in v=1 and v=3 remain backward peaked as at the lower collision energy, but the angular distribution for product in v=2 is very unusual. The intensity remains high as the center-of-mass scattering angle decreases, and even shows a sideways peaked relative maximum. This is illustrated more clearly when the center-of-mass velocity distributions are plotted as a function of angle as in Fig. 9. The laboratory angular distribution also shows a significant change between the two energies. The structure due to strongly backward peaked HF(v=2) product at 2.00 kcal/mole is absent when the collison energy is raised to 3.17 kcal/mole. The dramatic change in the center-of-mass angular distribution of the product formed in v=2 at 3.17 kcal/mole compared to that at 2.00 kcal/mole is consistent with the reactive resonance phenomenon predicted by the three dimensional calculations of Redmon and Wyatt. In that study it was shown that the effect of resonances in the F + H_2 reaction appears as a change in the dependence of the reaction probability on reactant orbital angular momentum. The reaction probability for forming HF(v=3) is maximum at zero orbital angular momentum, independent of collision energy, while the probability for forming HF(v=2) peaks at non-zero orbital angular momentum for a collision energy around 3 kcal/mole, and the orbital angular momentum corresponding to the peak reaction probability changes as a function of collision energy. Since the orbital angular momentum is determined by the impact parameter for the collision, the reaction probability for forming HF(v=2) peaks at a non-zero impact parameter for a collision energy around 3 kcal/mole. Products formed from a larger impact parameter collision should scatter at smaller center-of-mass angles, so our experimentally observed results at 3.17 kcal/mole are in qualitative agreement with what one would expect from the reactive resonance in $F + H_2 \rightarrow HF(v=2) + H$ predicted in the calculation of Redmon and Wyatt.

In the laboratory angular distribution of DF product from the reaction $F + D_2 \rightarrow DF + D$ at 4.51 kcal/mole (Fig. 4), there is another surprising feature: a sharp peak appears at the laboratory angle corresponding to forward scattering of DF(v=4). Such a sharp forward product peak has not been observed previously in a reaction which does not form a long lived complex. But just as the extent of the polarization of product orbital angular momentum determines the sharpness of forward-backward peaks in angular distributions of products from a long lived complex, the strong forward peaking of DF(v=4) indicates a strong correlation between the final orbital angular momentum of the product and the initial orbital angular momentum. The reliability of the angular momentum decoupling calculations carried out on $F + H_2 \rightarrow HF + H$ by Redmon and Wyatt depend on the extent of this correlation. For the F + H_2 and F + D_2 reactions, this approximation might appear to be valid. Once again, this strong correlation of final and initial angular

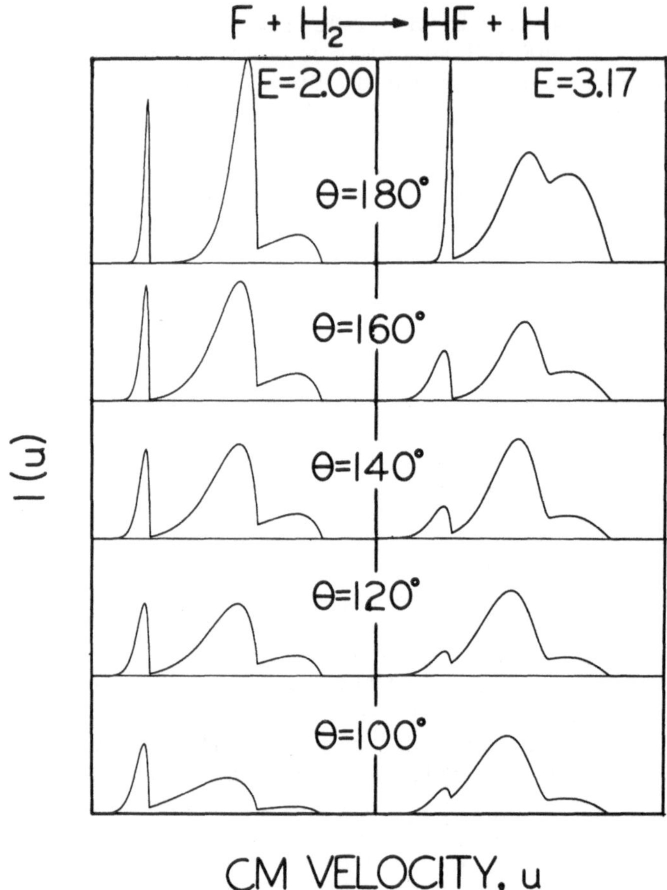

Fig. 9. Center-of-mass distributions at indicated angles of HF product from $F + H_2$ reaction at (a) 2.00 kcal/mole and (b) 3.17 kcal/mole.

momentum might be the consequence of the nature of the potential energy surface for these reactions, which has a bending angle dependent barrier which is minimum at the collinear configuration.

CONCLUSION

The experimental results presented in this paper are examples of what can be measured at present in the laboratory by the crossed beams method. The experimental resolution could be further improved in the future, but the incentive for future effort toward carrying out a better experiment certainly depends on the advancement of theoretical calculations, making possible a more quantitative comparison between theory and experimental results.

ACKNOWLEDGEMENT

This work was supported by the Division of Chemical Sciences, Office of Basic Energy Sciences, U.S. Department of Energy under contract No. W-7405-Eng-48.

REFERENCES

1. Kupperman, A., and Schatz, G.C.: 1975, J. Chem. Phys. 62, 2502.
2. Bender, C.F., Pearson, P.K., O'Neil, S.V., and Schaefer, III, H.F.: 1972, J. Chem. Phys. 56, 4626.
3. Muckerman, J.T.: 1971, J. Chem. Phys. 54, 1155.
4. Wilkins, R.L.: 1973, J. Chem. Phys. 58, 3038.
5. Schatz, G.C., Bowman, J.M., and Kupperman, A.: 1973, J. Chem. Phys. 58, 4023.
6. Latham, S.L., McNutt, J.F., Wyatt, R.E., and Redmon, M.J.: 1978, J. Chem. Phys. 69, 3746.
7. Redmon, M.J., and Wyatt, R.E.: 1979, Chem. Phys. Lett. 63, 209.
8. Parker, J.H., and Pimentel, G.C.: 1969, J. Chem. Phys. 51, 91; Coombe, R.D., and Pimentel, G.C.: 1973, J. Chem. Phys. 59, 251.
9. Polanyi, J.C., and Woodall, K.B.: 1972, J. Chem. Phys. 57, 1574.
10. Schafer, T.P., Siska, P.E., Parson, J.M., Tully, F.P., Wong, Y.C., and Lee, Y.T.: 1970, J. Chem. Phys. 53, 3385.
11. Lee, Y.T., McDonald, J.D., LeBreton, P.R., and Herschbach, D.R.: 1969, Rev. Sci. Instr. 40, 455.
12. Semco Instruments Incorporated, 11505 Vanowen Street, North Hollywood, California 91605 USA
13. Beam Dynamics Corporation, 623 East 57th Steet, Minneapolis, Minnesota 55417 USA.
14. James Martin Farrar, Ph.D. dissertation, The University of Chicago, Chicago, Illinois, 1974.
15. Hans-Dieter Meyer, Diplomarbeit, Max-Planck-Institut für Strömungsforschung, Göttingen, West Germany, 1974.

INTRAMOLECULAR DYNAMICS: REGULAR AND STOCHASTIC VIBRATIONAL STATES OF MOLECULES

R. A. Marcus
Noyes Laboratory of Chemical Physics, California Institute of Technology, Pasadena, California 91125 U.S.A.

ABSTRACT

The nature of molecular vibrational states at low and high energies has been studied both classically and quantum mechanically. At low energies the classical vibrational motion is highly "regular" (quasi-periodic) in the usual harmonic normal mode regime. Interestingly enough, it can still remain so when the motion is anharmonic. However, at high enough vibrational energies numerical calculations indicate that the classical mechanical motion usually becomes chaotic (frequently termed "stochastic") although still deterministic, of course. The corresponding quantum mechanical behavior is discussed using semi-classical ideas, and a method of calculating eigenvalues in the quasi-periodic regime is described.

A definition of quantum stochasticity is proposed in terms of overlapping avoided crossings in quantum mechanical eigenvalue versus perturbation parameter plots. Some implications for phenomena such as intramolecular relaxation, spectra, unimolecular reactions, and infra-red multiphoton dissociation of molecules are described.

INTRODUCTION

In the last twenty or so years, a "new phenomenon" has appeared in classical mechanics[1]. In this paper we describe this phenomenon and what we believe to be the analogous behavior in quantum mechanics. In the process we shall use some semiclassical results obtained by our research group[2] and indicate some of the implications for chemical behavior, as in intramolecular vibrational relaxation, unimolecular reactions, and infrared multiphoton dissociation of molecules.

Around the turn of the century, it was thought, we recall, that all was understood in physics; but then came quantum phenomena. However, not all was understood even in classical physics. In particular, in the classical mechanics of three-body systems in celestial mechanics,

K. Fukui and B. Pullman (eds.), Horizons of Quantum Chemistry, 107–121.

perturbation series were available for treating the motion, but these series were not shown to converge ("small divisors" problem) and were valid only for a limited time interval rather than for infinite time. Thus, the question of global time stability of the motion of an isolated solar system was unsolved. (Would some planet ultimately go to infinity?)

In the 1950's and early 1960's the existence of long time stability of classical mechanical systems was suggested[3] and proven[4], under restricted conditions, in the form of the now famous Kolmogorov-Arnol'd-Moser (KAM) theorem[3,4]. This theorem is concerned with the effect of small perturbations on integrable Hamiltonian systems. For integrable systems there exist as many single-valued constants of the motion, variously termed "first" or "isolating" integrals of the motion, as the number of degrees of freedom (N). This results in all the trajectories being forever confined to manifolds with the topology of N-dimensional tori (doughnuts) embedded in the 2N-1 dimensional phase space energy shell of the (conservative) system[5]. In this case the so-called action-angle variables are defined for all initial conditions (i.e., the Hamilton-Jacobi equation has a global solution). The KAM theorem states that for a sufficiently small nonintegrable perturbation, almost all tori are preserved, albeit in slightly distorted form. This means that with the exception of a set of small measure, the classical trajectories still display (infinite) long time stability. By contrast, the trajectories that do not lie on tori tend to wander in a pseudo-random manner (the motion is still deterministic) over large portions of the energetically available phase space. The integrals of the motion are, for the most part at least, not single-valued (? infinitely many sheets) and do not confine the trajectory the way tori do[6].

At low enough energies for weakly perturbed integrable systems such as coupled nonlinear oscillators, almost all of the trajectories lie on tori, while increasingly, in some transition region at higher energies, numerical calculations show that they tend to become of the wandering kind[1]. The two different classes of motion are often referred to as quasi-periodic and "stochastic" (or chaotic), respectively. Small changes in initial conditions can cause the trajectories to change from quasi-periodic to stochastic behavior[6].

CLASSICAL BEHAVIOR

A typical Hamiltonian for a pair of anharmonically coupled oscillators coordinates x and y and momenta, p_x and p_y, in a molecule is given by

$$H = \frac{1}{2}\left(p_x^2 + p_y^2 + \omega_x^2 x^2 + \omega_y^2 y^2\right) + \lambda x\left(y^2 + \eta x^2\right) \tag{1}$$

(This Hamiltonian is also related[7] to that for a star in the vicinity of the galactic plane in an axially symmetric galaxy[7,8]. The non-statistical velocity distribution of stars in a galaxy pointed to long time stability and restricted nature of the motion[7,8]. Thus, the dynamics based on (1) have been extensively investigated in the astronomy and related literature.)

When Hamilton's equations of motion for these coordinates and momenta are integrated numerically using a computer, one obtains a trajectory such as that in Fig. 1[9].

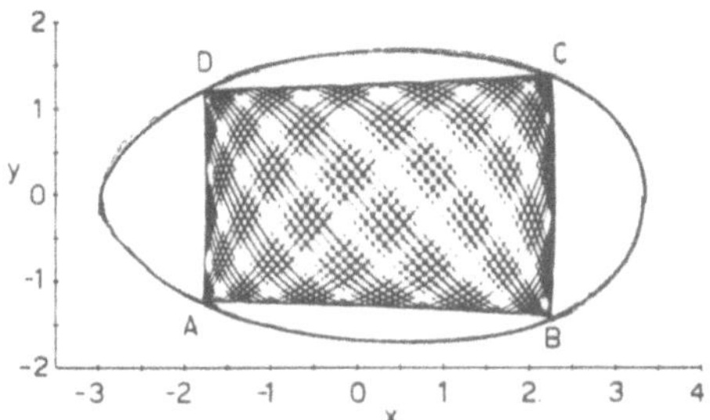

Figure 1. A trajectory for Hamiltonian (1), coordinate space xy, for the case ω_x and ω_y incommensurable.

The ellipse is the curve on which the potential energy equals the total energy and so the trajectory must lie in a region bounded by that curve. One sees, however, that the trajectory occupies a much more restricted region than the energetically accessible one. One sees, too, that the amplitude of the x-motion is approximately independent of y, and vice versa, and so there is relatively little energy interchange between the oscillators. The plot in Fig. 1 is a projection of the invariant torus in phase space onto the xy plane.

Another example is given in Fig. 2, now for the case that the unperturbed frequencies ω_x and ω_y are equal.

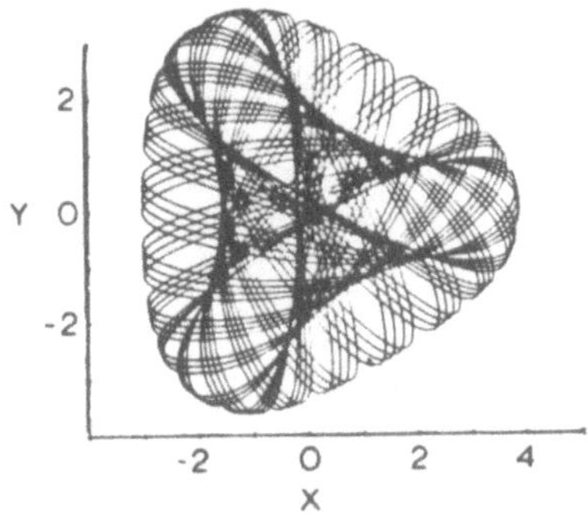

Figure 2. Same as Fig. 1 but for $\omega_x = \omega_y$.

The motion is again quasi-periodic at the low energy involved. Now, however[10], because of the "resonance" between the two oscillators, there is seen to be extensive interchange of energy: sometimes the motion is largely along the x direction, sometimes largely along the y one, and sometimes in between, during the course of the motion. Similarly, in the case of a 1:2 ratio of frequencies (Fermi resonance)[11] there is a fairly extensive interchange of energy, as in Fig. 3.

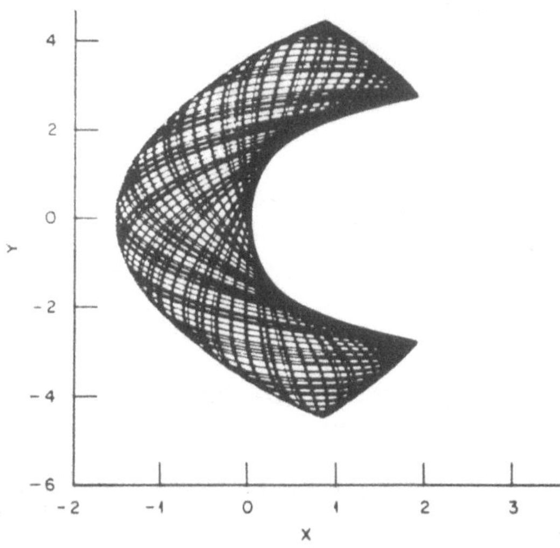

Figure 3. Same as Fig. 1 but for $\omega_x = 2\omega_y$.

A different plot is that of a Poincaré surface of section[12], e.g., a plot of p_x vs. x recorded each time the trajectory crosses the y = 0 axis with a (say) positive value of p_y, as in Fig. 4[9].

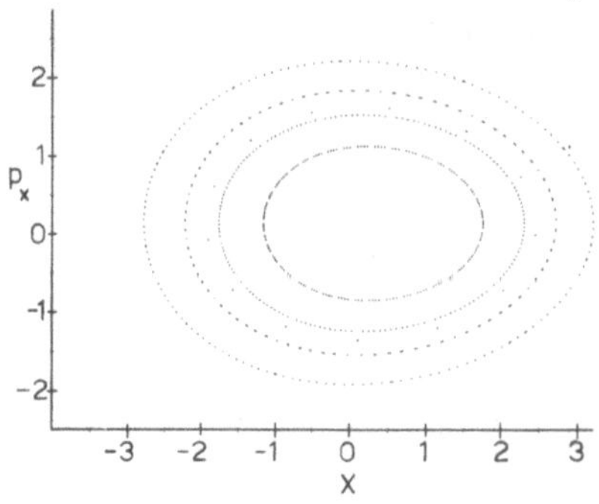

Figure 4. Poincaré surfaces of section for several trajectories, each a box-like one analogous to Fig. 1.

Here, the plot is given for a number of different trajectories at the same energy, each one giving rise to an ellipse-like figure. Related patterns arise in the 1:1 and 2:1 resonant systems, and here it has been convenient to introduce surfaces of section using curvilinear coordinates, patterned from the shapes of the regions in xy space covered by the trajectories (e.g., polar coordinates for the 1:1 resonant case and parabolic coordinates when ω_x/ω_y equals 1:2)[10,11].

The connection of these diagrams for the quasi-periodic regime with quantum mechanics is the following. Many years ago Einstein[13] pointed out that to quantize a dynamically nonseparable system, it was necessary to find the canonical invariants $\oint \underset{\sim}{p}\cdot d\underset{\sim}{r}$ for the system, when they exist, and set them equal to nh [or now, really $(n_i + \frac{1}{2})h$, in the case of oscillators], each cyclic integral in the phase space being over a topologically independent path C_i in the phase space. Those for the Fig. 1 have been described elsewhere[9] and are, for example, the same as or equivalent to the y = 0 and x = 0 Poincaré surfaces of section. Those for Figs. 2 and 3 were obtained from curvilinear[10,11] surfaces of section. The number of independent integrals equals the number of coordinates (N, say). The allowed energies of the system are those for which the numbers $(n_1, \ldots, n_N)$ are integers. Using such semiclassical concepts, derived now from multidimensional WKB type arguments[14,2a], one obtains in this way good agreement between eigenvalues for the Hamiltonian (1) calculated quantum mechanically from a large variational basis set and those calculated semiclassically[9-11]. Semiclassical eigenvalues were calculated for the first time for two or more dimensional nonseparable systems with smoothly varying potentials in Refs. 15 and 9. More recently, a variety of methods have been developed[16].

The wave function is large and oscillatory in the shaded region in Figs. 1 to 3, and decreases exponentially outside that region in accordance with multidimensional semiclassical theory. The boundaries serve as caustics for the wave function. There is also a regular nodal pattern (where we have examined it) in this quasi-periodic regime[11].

From a classical trajectory a correlation function, e.g., for x(t) and y(t), has been calculated, and from it by a Fourier transform a power spectrum was obtained[17]. The spectrum consisted of several lines. The positions of the spectral lines agreed well with the corresponding quantum mechanically calculated lines (i.e., with the differences of quantum mechanical eigenvalues)[17].

At high energies, a quite different behavior occurs. The trajectory now tends to occupy much of the energetically accessible space. Correspondingly, the Poincaré surface of section produced by a trajectory tends to be a shotgun pattern, as in Fig. 4 of Ref. 1a, and one can no longer evaluate $\oint \underset{\sim}{p}\cdot d\underset{\sim}{r}$ integrals ("action" variables) from these surfaces of section. Indeed, in a truly stochastic regime, the only isolating integral of the motion is the total energy (and, where applicable, the angular and total momentum). The N $\oint \underset{\sim}{p}\cdot d\underset{\sim}{r}$ integrals appear to no longer exist. Thus, it has not been possible, as yet, to obtain semiclassical eigenvalues from classical trajectories. A method which gives the value to within one quantum state, for the system studied, from evaluations of volume of classical phase space has been devised[2f].

The classical spectrum, obtained from the correlation function, now consists of numerous lines, giving rise to a "band" near the fundamental frequency as in Fig. 6 of Ref. 2c. The quantum mechanical wave function in this region appears to have an irregular nodal pattern where studied[2e,11,18].

QUANTUM STOCHASTICITY

In a quasi-periodic energy regime, we have noted, the wave function is fairly localized, whereas in the stochastic energy regime it is expected to be more delocalized, largely over the microcanonically classically accessible region. On that basis we suggested[2e] that for a quantum state corresponding to a truly stochastic regime the quantum mechanical average of each dynamical quantity A(p,r) would approximately equal the microcanonical average at that energy (provided A does not weigh heavily the classically forbidden regions). A possible mechanism for the onset of this statistical nature of the wave function in the stochastic regime is the following.

For some unperturbed Hamitonian H_o which is integrable, there tend to be regular sequences of the energy levels. With increase of some perturbation parameter, some isolated energy levels may tend to come near each other and undergo an avoided crossing as in Fig. 5.

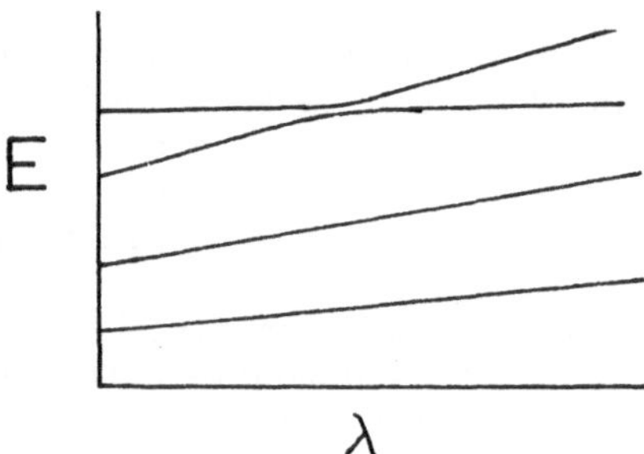

Figure 5. Example of an isolated avoided crossing in a plot of energy eigenvalues E versus a perturbation parameter λ.

An example of this behavior for the Hamiltonian (1) is given in Ref. 2f. In the vicinity of the avoided crossing each of the two perturbed wave functions takes on the character of both 'zeroth order wave functions' at that λ. Thus, the nodal pattern is more complex than before (in contrast to an actual crossing, which is not expected to change the nodal pattern). This isolated avoided crossing does not in itself yet constitute "quantum stochasticity." Rather, it represents the quantum analog of an isolated classical resonance. A classical Hamiltonian with an isolated resonance is integrable[1].

Classically, an isolated resonance occurs where, for some set of initial conditions, the characteristic frequencies $\bar{\omega}_i$ of the motion (which depend on the perturbation parameter and on the action variables and hence on the initial conditions) become commensurable, i.e., when there are positive and negative integers m_i such that

$$m_1\bar{\omega}_1 + \ldots + m_r\bar{\omega}_r = 0. \tag{2}$$

One can then define new action variables, and thereby, new frequencies $\bar{\omega}_i'$, such that at this resonance one of the new frequencies is zero. Since frequencies correspond semiclassically to differences of quantum mechanical energy eigenvalues, the corresponding 'zeroth order eigenvalues' versus perturbation parameter plots should intersect.

We anticipate that quantum stochasticity will begin in some energy region when many eigenvalue curves undergo avoided crossings there[2f], as depicted schematically for the crossing of three such curves in Fig. 6.

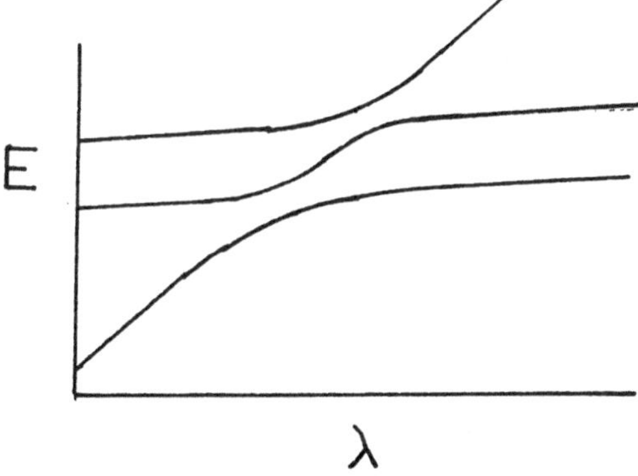

Figure 6. Example of overlapping avoided crossings in the plot for Fig. 5.

Now each of the wave functions contains the character of each of the three 'zeroth order wave functions', for λ in the vicinity of the crossings, and so has begun to take on a statistical character. When many such overlapping avoided crossings occur in the same region of parameter space and energy, something more likely to happen at high energies because in part of the higher density of states, each of the relevant wave functions has become "stochastic," and the average of any dynamical quantity in that state should become more nearly equal to the classical microcanonical average at that energy. At present, we are investigating the dynamics of this behavior[19]. The present definition is less basis-set dependent that one based on the projection of the exact wave function onto those for H_o.

We believe that 'overlapping avoided crossing' is a quantum analog of Chirikov's "overlapping" of classical mechanical resonances[20]. Chirikov has used this overlapping as a tool for predicting onset of stochasticity[21] in classical mechanical systems. However, the presence of "quantum stochasticity" as we have defined it, i.e., the presence of overlapping avoided crossings, is not necessarily implied by the presence of a corresponding classical stochasticity. This is due to the finite size of $\hbar$, which can lead to a sufficiently large energy level spacing such that avoided crossings are relatively rare occurrences (see, for example, Ref. 2f). However, in the semiclassical limit $\hbar \to 0$ there could well be a close connection between the classical stochasticity and our definition of quantum stochasticity. In this case our "quantum stochastic states" would be closely related to the "irregular states" postulated by Percival[19a].

CONSEQUENCES OF QUANTUM STOCHASTICITY

One main difference in the two regimes lies in the spectrum. In the quasi-periodic regime it should be decipherable, in principle, into regular sequences (interlacing ones when there are several degrees of freedom). For example, for a set of N coupled oscillators these allowed frequencies are differences of eigenvalues

$$E_{n_1 \ldots n_N} = \sum n_i \, \omega_i(\underset{\sim}{n}) \, \hbar \; + \text{constant} \tag{3}$$

where the frequencies ω_i, due to anharmonicity and to anharmonic coupling, are slowly varying functions of $n_1 \ldots n_N$. In the stochastic region, due to overlapping avoided crossings at the given value of the perturbation parameter, the spacings of eigenvalues become more complicated and "erratic," creating an irregular spectrum. Isolated avoided crossings would cause isolated irregularities in the spectrum. Because of the large number of vibrational states available at higher energies, it may be difficult to sort out any regular sequences (3) even if they occurred there.

A second effect of quantum stochasticity on a vibrational spectrum is to distribute the infrared oscillator strength in a particular frequency range over many true eigenstates and so "broaden" the spectrum. For example, if most of this oscillator strength came from a particular coordinate (say), it would, if the final vibrational quantum states were in the stochastic regime, be distributed over many final true vibrational eigenstates and so give rise to a broadened spectrum (cf. 19a). Some distribution of oscillator strength could also occur even if the final eigenstates were in the quasi-periodic regime, and so cause some (but less) broadening. It remains for detailed calculations for individual cases to compare the relative amount of broadening.

Examples of broadened spectra for low and high overtones are found in CH bonds in aromatics[22] (usually complicated by presence of several CH bonds) and C=O bonds in ketones[23]. The detailed interpretation of the width of a high overtone spectrum is of particular interest. It may be due to "energy relaxation" or to a "pure vibrational dephasing" mechanism, or to both[24]. "Pump and probe" experiments should distinguish the two. In the latter one excites ("pumps") one absorption band and monitors the intensities of other bands. Such experiments are, for intramolecular vibrational relaxation, at or almost at hand.

When the width is predominantly due to an "energy relaxation mechanism," there is a close relationship between the width and the relaxation time for an upper state population formed from a pulse excitation of the entire band. When the spectral width is much larger than can be explained in this way and when it is not due to the presence of overlapping absorption bands ("inhomogeneous broadening"), one has, instead, a "pure vibrational dephasing mechanism."

To illustrate the relation of the two mechanisms to exact final eigenstates, it is convenient to consider first a very simple model. Here, one coordinate x_1 is assumed to carry the oscillator strength for the band, via a dipole matrix element over two zeroth order wave functions $\phi(x_1)$ and $\phi^*(x_1)$ for the lower and upper states, respectively. We let $\phi_m(x_1)$ denote the remaining zeroth order wave functions for x_1 and let $x_2 \ldots x_N$ be the remaining coordinates, with wave functions $\psi_n(x_2 \ldots x_N)$. We consider for simplicity the excitation from a single vibrational state, e.g., the ground state of the system.

The (normalized) "exact" wave function for the entire molecule in one of its upper vibrational states ψ_f in the band is then, for this model,

$$\psi_f(x_1 \ldots x_N) = \phi \sum_n a_n \psi_n + \phi^* \sum_n b_n \psi_n + \sum_{mn} c_{mn} \phi_m \psi_m \,. \tag{4}$$

There can be many ψ_f's, each differing in values of the a's, b's and c's. If, as one limit, the underlying perturbed classical motion for the final states were integrable and were as in Fig. 1, with x denoting x_1 and y denoting collectively $x_2 \ldots x_N$, each ψ_f would have only one dominant term and would be quasi-periodic: only a single $|b_n|$ in it is near unity, all remaining coefficients would be small, or, in the case of an isolated avoided crossing at most two of $|b_n|$'s in it would be appreciably large. If a ψ_f quantum state in this model were, instead, stochastic, none of the $|a_n|$'s $|b_n|$'s and $|c_{mn}|$'s would be typically large.

A pure vibrational dephasing mechanism would prevail if all of the "exact" states ψ_f containing ϕ^* did not contain a significant contribution from ϕ and from the ϕ_m's, i.e., if the exact wave functions ψ_f were of the form $\phi^* \sum_n b_n\psi_n$. In this case the ψ_f's are, at most, statistical only in their description of the $(x_2 \ldots x_n)$ motion. Other things being equal one expects the band width to be less when the ψ_f's are quasi-periodic than when they are statistical.

An energy relaxation mechanism prevails when all of the $|a_n|$'s, $|b_n|$'s and $|c_{mn}|$'s are small for the upper states. Then, the final quantum states are stochastic. (In more complicated situations there can also be intrinsic internal resonances which can contribute to the nature of the relaxation.) One model frequently used for other types of systems is the "discrete state in a continuum"[25,26] and is a particular case of Eq. (4). Here, the (unperturbed) "discrete state" is a single $\phi^* \psi_{n_0}$ and the continuum of states is $\phi \Sigma a_n \psi_n$, each such state with a different set of values of the a_n's. The "exact" eigenstates are then linear combinations of the discrete state and the continuum state, and so correspond to (4) with all b_n's but one equal to zero, and all c_{mn}'s equal to zero.

In contrast, when x_1 does not nearly parallel the x (or y) axis in Fig. 1, an "energy relaxation" in the x_1 coordinate, a relaxation defined operationally, would be observed even when the ψ_f's are quasi-periodic. (Cf. an analogous relaxation in an integrable system, the Toda lattice[27].)

In the above we have omitted for brevity the case where intrinsic internal resonances occur in H_0, as in Figs. 2 and 3. They permit periodic energy exchange instead of or superimposed on an "irreversible" one.

One also expects differences in behavior of a molecule undergoing optical excitation, according as it is excited to a quasi-periodic or a stochastic vibrational quantum state of an upper electronic state. In the former case, the subsequent behavior would depend very much on which vibrational state it is excited to, varying markedly from state to state. For example, if the excitation were to some given metastable electronic state and to various vibrational states, the subsequent fluorescence or predissociation behavior of this electronically and vibrationally excited molecule could depend significantly on that vibrational state. If the upper vibrational state were, instead, of a stochastic nature, then because of its statistical character, less dependence on the individual upper vibrational state would be anticipated.

Again, if a particular collision largely excited some terminal bond in a molecule and transferred enough energy to place the molecule in the stochastic regime, one could again form a "discrete state in a continuum" system. If the range of energies of excited states exceeded the width of states coupled to the discrete state, one would have a phenonemon of intramolecular energy relaxation. In the usual theory of unimolecular reactions (RRKM theory)[28] rapid intramolecular "randomization" is assumed so that the subsequent behavior of the molecule depends on its energy (and angular momentum) but not (or not noticeably) on any other integrals of the motion. A significant dependence on the latter could give deviations from RRKM theory[29]. A correlation function approach to time scales for energy relaxation is discussed in Ref. 30.

In the phenomenon of infrared multiphoton dissociation of a molecule current theoretical work assumes that these are two energy regimes of the molecule, with quite different properties--a discrete state regime and a quasi-continuum[31]. In the former a coherent absorption of the infrared radiation is assumed from state to state, while in the second regime rapid intramolecular randomization is frequently supposed. These two regimes may correspond largely to the present quasi-periodic and stochastic regimes, although the laser width is also involved in the definition. An example of a classical mechanical trajectory for a CD_3Cl molecule undergoing infrared multiphoton dissocation is given in Fig. 3 of Ref. 32, in which the energy absorbed by the molecule is plotted versus time. The "induction period" may correspond to a response of the molecule in its initial quasi-periodic regime with absorption and emission occurring. It is followed by a more cumulative increase of energy, perhaps when the molecule reaches the stochastic regime and can more readily undergo an intramolecular relaxation. A more detailed investigation of this phenomenon is underway[33].

SUMMARY

We have summarized briefly the "new" phenomenon in classical mechanics. Recent symposia and reviews attest to its widespread interest. It also arises in many other systems, as in a transition from laminar to turbulent flow. We have also indicated what we believe its counterpart in quantum mechanics to be and some implications for various experimental systems. Because of the detailed information becoming available through the use of lasers and other techniques, the subject is one of considerable current interest.

ACKNOWLEDGMENT

It is a pleasure to acknowledge the contributions of my coworkers, Drs. M. Tabor, D. M. Noid, M. L. Koszykowski, W.-K. Liu, and R. Ramaswamy and the support of this research by the National Science Foundation. I am particularly indebted to Dr. Tabor for his critical reading of the manuscript and for his very helpful advice. Contribution No. 6145 from the California Institute of Technology.

REFERENCES

1. E.g. (a) Ford, J.: 1973, Adv. Chem. Phys. 24, p. 155; Ford, J.: 1975, in "Fundamental Problems in Statistical Mechanics," Cohen, E.D.G. ed. (North Holland, Amsterdam), vol. 3; Tabor, M.: 1980, Adv. Chem. Phys. (in press); P. Brumer, ibid., and the many references cited in these review articles; (b) Articles in "Stochastic Behavior in Classical and Quantum Systems, Volta

Memorial Conference, Como, 1977," Casati, G. and Ford, J. eds. (Springer Verlag, New York); (c) Berry, M. V.: 1978, Am. Inst. Phys. Conf. Proc. 46, "Topics in Nonlinear Dynamics," Jorna, S. ed., p. 16; for definitions of integrability, etc. see Helleman, R. G., ibid., p. 400.

2. E. g. (a) Marcus, R. A., Noid, D. W., and Koszykowski, M. L.: 1978, in "Advances in Laser Chemistry," Zewail, A. H., ed. (Springer Verlag, New York), p. 298; (c) Marcus, R. A., Noid, D. W., and Koszykowski, M. L.: 1979, in ref. 1b, p. 298; (d) Noid, D. W., Koszykowski, M. L., and Marcus, R. A.: 1979, J. Chem. Phys. 80, p. 2864 and refs. cited therein; (e) Marcus, R. A.: 1979, in "Extended Abstracts, 27th Annual Conference on Mass Spectrometry and Allied Topics, Seattle, Washington, June 3-8"; (f) Noid, D. W., Koszykowski, M. L., Tabor, M., and Marcus, R. A. (to be submitted), "Properties of Vibrational Levels in the Quasi-Periodic and Stochastic Regimes."

3. Kolmogorov, A. N.: 1954, Dokl. Akad. Nauk. SSSR, 98, p. 527; English translation: Ref. 1b, p. 51.

4. Arnol'd, V. I.: 1963, Russian Math. Surveys, 18, p. 85; Moser, J.: 1962, Nachr. Akad. Wiss. Göttingen, II Math. Physik Kl., No. 1, p. 1; Arnol'd, V. I. and Avez, A.: 1968, "Ergodic Problems of Classical Mechanics" (Benjamin, New York); Arnol'd, V. I.: 1978, "Mathematical Methods of Classical Mechanics, Graduate Texts in Mathematics (Springer Verlag, New York). The theorem is for non-degenerate oscillators, but applies also when there are two oscillators which, in zeroth order, are degenerate (cf. Arnol'd, V. I., loc cit. and Ref. 1a; Moser, J.: 1973, "Stable and Random Motions in Dynamical Systems" (Princeton Univ. Press, Princeton).

5. When total angular momentum and total momentum are also conserved quantities, the 2N-1 becomes 2N-5.

6. Thus, in this case, the only isolating integrals of the motion are the energy and, where applicable, the angular and linear momentum.

7. Contopoulos, G.: 1960. Z. Astrophys. 49, p. 273; Hori, G.: 1967, Publ. Astron. Soc. Japan 19, p. 229.

8. E. g., Ollongren, A.: 1962, Bull. Astron. Inst. Neth. 16, p. 241; Hénon, M. and Heiles, C.: 1964, Astron. J. 69, p. 73.

9. E. g., Noid, D. W. and Marcus, R. A.: 1975, J. Chem. Phys. 62, p. 2119.

10. E. g., Noid, D. W. and Marcus, R. A.: 1977, J. Chem. Phys. 67, p. 559.

11. E. g., Noid, D. W., Koszykowski, M. L., and Marcus, R. A.: 1979, J. Chem. Phys. 80, 2864.

12. Poincaré, H.: 1892, "Les Méthodes Nouvelles de la Méchanique Céleste," 1957 (Dover, New York), vol. 1; ibid.: 1976, English transl. NASA, Washington, D.C., vol. 3, chap. 27.

13. Einstein, A.: 1917, Verh. Dtsch. Phys. Ges. 19, p. 82.

14. Keller, J.: 1958. Ann. Phys. (N. Y.) 4, p. 180.

15. Eastes, W. and Marcus, R. A.: 1974, J. Chem. Phys. 61, p. 4301.

16. E. g., Percival, I. C. and Pomphrey, N.: 1976, Mol. Phys. 31, p. 917; Chapman, S., Garrett, B., and Miller, W. H.: 1976: J. Chem. Phys. 64, p. 502 ; Sorbie, K. S. and Handy, N. C.: 1976, Mol. Phys. 32, p. 1327; Sorbie, K. S.: 1976, ibid. 32, p. 1577; Handy, N. C., Colwell, S. M., and Miller, W. H.: 1977, Faraday Discuss, Chem. Soc. 62, p. 29; Sorbie,K. and Handy.N.: 1977, Mol. Phys. 33, p. 1319; Delos, J. B. and Swimm, R. T.: 1977, Chem. Phys. Lett. 47, p. 76; Jaffé, C. and Reinhardt, W. P.: 1979, J. Chem. Phys. 71, p. 1862; Percival, I. C.: 1979, ref. 2b, p. 259; Colwell, S. M., Handy, N. C. and Miller, W. H.: 1979, ref. 2b, p. 299.

17. Noid, D. W., Koszykowski, M. L., and Marcus, R. A.: 1977, J. Chem. Phys. 67, 404; Hansel, K. D.: 1978, Chem. Phys. 53, p. 35.

18. J. D. McDonald, unpublished results.

19. Other viewpoints for quantum stochasticity are found in (a) Percival, I. C.: 1973, J. Phys. B6, p. 1229; Pomphrey, N.: 1974, ibid. 7, p. 1909; (b) Pechukas, P.: 1972, J. Chem. Phys. 57, 5577; review by Tabor, M., ref. 1a; Nordholm, K. S. J. and Rice, S. A.: 1974, J. Chem. Phys. 61, p. 203, 768; Berry, M. V. and Tabor, M.: 1976, Proc. Roy. Soc. Lond. A349, p. 101; Percival, I. C.: 1977, Adv. Chem. Phys. 36, p. 1; Berry, M. V.: 1977, Phil. Trans. Roy. Soc. 287, p. 237; Heller, E.: 1979, Chem. Phys. Letters 60, p. 338; Heller, E., "Quantum Intermolecular Dynamics; Crtieria for Stochastic and Non-Stochastic Flows" (preprint); Stratt, R. M., Handy, N. C., and Miller, W. H.: 1979, J. Chem. Phys. 71, p. 3311; Shuryak, E. V.: 1976, Sov. Phys. JETP 44, 1070.

20. Zaslavskii, B. M. and Chirikov, B. V.: 1972, Sov. Phys. Usp. 14, p. 549; Chirikov, B. V.: 1979, Phys. Reps. 52, p. 265; cf. Ford, J. and Lundsford, G. H.: 1970, Phys. Rev. A1, p. 59.

21. For other criteria proposed for the onset of classical stochasticity, e.g., of the Toda-Brumer and Mo methods, see ref. 1a.

22. For review see articles by Reddy, K. V., Bray, R. G., and Berry, M. J., and by Albrecht, A. C.: 1978, in "Advances in Laser Chemistry," Zewail, A. H., ed. (Springer Verlag, New York), pp. 48, 235, respectively; Perry, J. W. and Zewail, A. H.,: 1979, J. Chem. Phys. 65, p. 31.

23. Smith, D. D. and Zewail, A. H.: 1979, J. Chem. Phys. 71, p. 540.

24. Cf. Jones, K. E. and Zewail, A. H., ref. 22, p. 196; Diestler, D. J.: 1976, Chem. Phys. Lett. 39, p. 39; Diestler, D. J.: 1976, Mol. Phys. 32, p. 1091; Oxtoby, E. W. and Rice, S. A.: 1976, Chem. Phys. Lett. 42, p. 1; Madden, P. A. and Lynden-Bell, R. M.: 1976, ibid. 38, p. 163. A molecular description of collisionally-induced T_1 and T_2 relaxation times is given in Liu, W. -K. and Marcus, R. A.: 1975, J. Chem. Phys. 63, pp. 272, 290.

25. Fano, U.: 1961, Phys. Rev. 124, p. 1866; cf. Harris, R. A.: 1963, J. Chem. Phys. 39, p. 978.

26. For discussions of spectra and relaxation cf. Rhodes, W.: 1969, J. Chem. Phys. 50, p. 2885; Chock, D. P., Jortner, J., and Rice, S. A.: 1967, ibid. 49, p. 610; Langhoff, C. A. and Robinson, G. W.: 1973, Mol. Phys. 26, p. 249; Tric. C.: 1973, Chem. Phys. Lett. 21, p. 83; Tric. C.: 1974, Chem. Phys. 14, p. 189, and references cited therein; Freed, K. F.: 1974, Topics Appl. Phys. 15, p. 23.

27. Ford, J., Stoddard, S. D., and Turner, J. S.: 1973, Progr. Theoret. Phys. 50, 1947; cf. Hénon, M.: 1974, Phys. Rev. B 9, p. 1921; Flashka, ibid., p. 1924.

28. Marcus, R. A.: 1952, J. Chem. Phys. 20, 359; 1965, ibid. 43, p. 2658; 1970, ibid. 52, 1018; Robinson, P. J. and Holbrook, K. A.: 1972, "Unimolecular Reactions" (Wiley, New York); Forst, W.: 1973, "Theory of Unimolecular Reactions" (Academic Press, New York); Spicer, L. D. and Rabinovitch, B. S.: 1970, Ann. Rev. Phys. Chem. 21, p. 349.

29. Marcus, R. A.: 1977, Ber. Bunsenges. Phys. Chem. 81, p. 190.

30. Tabor, M., Marcus, R. A., Koszykowski, M. L. and Noid, D. W. "On Correlation Functions and the Onset of Chaotic Motion" (to be submitted).

31. E. g., review of Bloembergen, N. and Yablonovitch, E.: 1978, Physics Today, 31, p. 23; Goodman, M. D., Stone, J., and Thiele, E.: 1979, in "Multiple-Photon Excitation and Dissociation of Polyatomic Molecules," Topics Curr. Phys., Cantrell, C. D., ed. (Springer Verlag, New York) (in press); Mukamel, S. and Jortner, J.: 1976, J. Chem. Phys. 65, p. 50-52; Grant, F. R., Coggiola, M. J., Lee, Y. T., Schultz, P. A., Sudbo, A. S., and Shen, Y. -R.:

1977, Chem. Phys. Lett. 52, p. 595; Quack, M.: 1979, Chem. Phys. Lett. 65, p. 140.

32. Noid, D. W., Kosykowski, M. L., Marcus, R. A., and McDonald, J. D.: 1977, Chem. Phys. Lett. 51, p. 540.

33. Ramaswamy, R. and Marcus, R. A. (unpublished).

INTERACTIONS BETWEEN CHIRAL MOLECULES : DISCRIMINATING INTERACTIONS

D.P. Craig
Research School of Chemistry, The Australian National University, Canberra, A.C.T., Australia 2600.

K. Fukui and B. Pullman (eds.), Horizons of Quantum Chemistry, 123–143.

I. EXPERIMENTAL BACKGROUND

I.1 Introduction

The force between two molecules or ions A and B, and the energy of their mutual coupling, is conventionally related to quantities such as electric charge, electric dipole moment, electric polarisability and short-range exchange repulsions. These quantities are the same for molecules and their optical isomers. The properties characteristic of the chirality of optically active systems, such as electronic transitions allowed to both electric and magnetic radiation fields, and special combinations of electric moments, are sources of intermolecular energy which are swamped by the contributions common to molecules and ions in general. Thus if *l*A and *d*A, and *l*B and *l*B, are the optical antipodes of molecules A and B, the interaction energy difference for the pairs (I.1)

$$\left.\begin{array}{l} lA - lB \\ lA - dB \end{array}\right\} \qquad (I.1)$$

is in all cases so far known a small fraction of the interaction common to the two members of the pair. In making such statements one is taking for granted that the comparison is made in a way keeping *l*A fixed, and generating *d*B from *l*B by a symmetry operation that keeps the intermolecular separation R the same, according to some prescription. The differences between the interactions (I.1) are said to arise from *chiral discrimination*. Some aspects to be described have been covered in part recently [1,2]. As will be seen, realistic comparison of the systems can often be made only after allowing the molecules to rotate freely, and averaging over all relative orientations.

Chiral molecules are molecules which exist in two (chemically identical) forms which cannot be mapped on one another by any symmetry operation. This is expressed in the condition that the covering group of the molecule cannot include an improper axis of rotation among its elements. In particular, the group cannot include inversion (improper rotation of zero angle) or reflection in a mirror plane (improper rotation of angle π). Proper rotations may occur: they map one chiral form on itself but not on its antipode. One form can be mapped on to the other only by operations that are disallowed, all of which have the effect of transforming a left- into a right-handed reference frame when treated as operating on the coordinate frame instead of on the molecule. The point groups in which chirality can occur are readily found to be $\tilde{C}_n$, $\tilde{D}_n$, $\tilde{T}$, $\tilde{O}$ and $\tilde{I}$. No molecular examples have been discovered in the last three higher symmetry groups. Among $\tilde{C}_n$ examples are included molecules lacking all symmetry such as those with a central atom (say carbon) attached to four atoms or groups, A, B, C, D all different (Fig. 1a). The $\tilde{D}_n$ groups are examplified (Fig. 1b) by the metal *tris*-chelate cations such as $[Co(ethylene\ diamine)_3]^{+++}$ which have the propeller-like D_3 symmetry, and by the distorted *spiro*-nonatetraene (1c) structure in $\tilde{D}_2$, believed [3] to be generated by Jahn-Teller distortion in an upper state from the normal $\tilde{D}_{2d}$ structure.

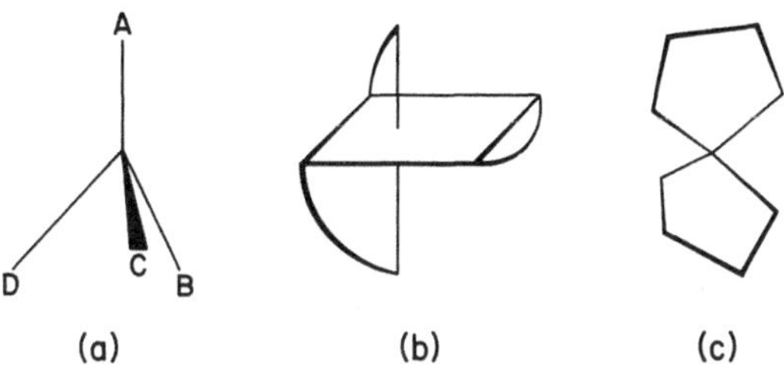

Figure 1. Examples of chiral molecules in $\underset{\sim}{C}_1$ $\underset{\sim}{D}_3$ and $\underset{\sim}{D}_2$

I.2 Experimental indications of chiral discrimination

The remarks in Sec. I.1 apply to small molecules in which electrical properties can be treated as possessing point sources. Historically however the first discoveries of discrimination were in biosystems so large that their properties must be distributed over a framework: much greater effects are then seen. Pasteur found in 1858 that the mould *penicillium glaucum*, in dilute solution of the salt formed between ammonium (NH_4^+) and equal amounts of the *l* and *d* forms of tartaric acid, attacked only the *d* form, being inert towards the *l*. Among the more recent examples there is the Pfeiffer effect, a case of which is the displacement of the equilibrium between (+) and (-) forms of the $\underset{\sim}{D}_3$ complex cation [(o-phenanthroline)$_3$Ni^{II}] by addition of the chiral (+) cinchoninium cation [4]. The Ni^{II} complex cation is optically labile, and the concentrations of its optical isomers change in response to their different interactions with the added chiral substance. The results may be expressed in thermodynamic terms, where the standard Gibbs free energy required to convert one optical form into the other goes from zero (in water) to 0.34 kJ mol^{-1} in a solution of (+) cinchoninium at 0.054 M.

Even in this example, the cinchoninium cation is too large to establish a basis for theoretical discussion. We quote one further result which approaches more closely to the small molecule case, due to Barnes, Backhouse, Dwyer and Gyarfas [5]. They measured the electrode potentials for the redox systems (+)-[Os(bipy)$_3$[$^{2+}$/(+)-[Os(bipy)$_3$]$^{3+}$ and the corresponding (-) systems in the presence of added chiral ions, as a function of the ionic strength. The effect of the added chiral ions is to stabilise the (+) or (-) systems differentially, causing a change from equality in the redox potentials. For added [Co(en$_3$)]$^{3+}$ cation, the redox potential difference amounted to 50 J mol^{-1} at ionic strength $I^{1/2}$ = 0.10. The osmium and cobalt ions carry the same charge, and should aggregate into ion pairs much less than would oppositely charged ions. Also they are much smaller and more compact than in most examples of discrimination. Both points are important in comparing experiment with the theory to be discussed.

I.3 Chiral discrimination by differential crystal packing

It is not surprising that many striking discriminations occur in crystalline solids. Individual crystals grown from a racemic solution of threonine consist of chirally pure (+) or (-) material, showing that such crystals are more stable than racemic crystals. In tartaric acid, racemic crystals are more stable than chirally pure crystals. The general explanation appears to be that the molecular surface contours within the *l*-*l* and *l*-*d* pairs enable more compact packing in one or the other, varying from case to case, leading to structures of different densities, and different average molecular separations. Since the strongest interactions, notably the dispersion interaction, change with distance as R^{-6} or faster, the variations of average separation can be the source of large packing energy differences, as shown in lattice energies, heats of sublimation and in heats of solution in various solvents. Such attractive interactions need not be discriminating in the technical sense; the differences arise from non-discriminating forces acting at different distances. The discrimination has its source in the very short range repulsions, which act as though the molecules had more or less hard surfaces, and impose different packing patterns. We do not deal with such cases in this paper.

As an illustration of discriminations in crystals we can take the compound 2,2,5,5-tetramethyl-3-hydroxypyrrolidine nitroxide (Fig. 2) studied by Chion *et al.* [6] in crystals of pure enantiomer and

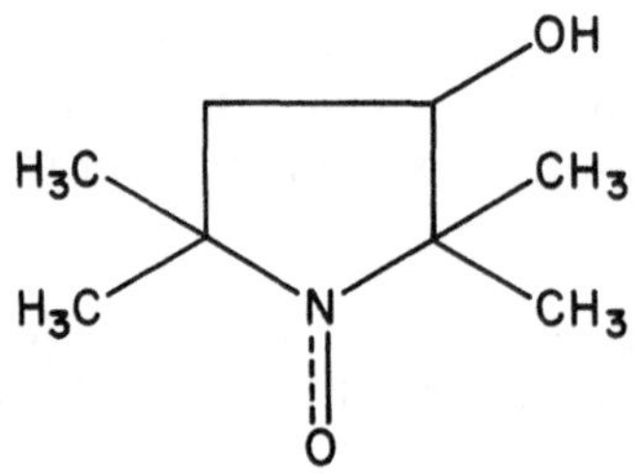

OH
H_3C CH_3
H_3C N CH_3
O

Figure 2. 2,2,5,5-tetramethyl-3-hydroxypyrrolidine

pseudo-racemate. The densities of the crystals are respectively 1.133 and 1.119 g cm^{-1}. On average the molecules in the pure enantiomer are closer than those of the racemic mixture by 0.4%. For that reason alone the dispersive binding in the pure enantiomer should exceed that in the pseudo-racemate by 0.3-0.5 kJ mol^{-1}, and this excess will be one of the terms in the differences between lattice energies. The overall energy balance does not necessarily make the more tightly packed crystal the more stable; in trans-1,2-dibromoacenaphthene for example the packing is denser in the active than in the racemic crystal (densities 1.52 and 1.38 g cm^{-3}), but the racemic form is a little more stable thermodynamically [7].

II. DISCRIMINATION IN THE DISPERSION INTERACTION

II.1 Discriminating interactions at ranges beyond contact distance

We now discuss discrimination at separation distances large enough compared with molecular dimensions to allow intermolecular forces to be treated as if they acted between points in the interacting bodies. Molecular properties, such as electric and magnetic moments, are lumped at these points as origins. In molecules with no symmetry there is no natural origin, and the choice, often the centre of mass, is arbitrary. In molecules of groups $\underset{\sim}{D}_2$ and $\underset{\sim}{D}_3$ there is a natural molecular origin where the two-fold axes intersect (Figs. 1c and 1b).

It has already been noted that the discriminating terms in the interaction energy are much less than non-discriminating terms. It is expected that in many cases the non-discriminating terms will impose a preferred relative orientation on the coupled molecule pair, and the discriminating terms will have to be evaluated at this orientation, or perhaps Boltzmann-weighted over a range of orientations. For ease of classification it is best to take first the case of molecules of which the orientations are to be randomly averaged over all orientations (isotropic averaging), and to describe as discriminating interactions any that give different energies for *l*A-*l*B and *l*A-*d*B pairs. According to this test only dispersion interactions, of the simple intermolecular couplings as usually understood, can be discriminating.

The dispersion interaction as usually understood is a pure electric second order coupling based on the Hamiltonian (II.1)

$$H = H_A + H_B + H_E$$

where

$$H_E = - \sum_{iq} \frac{Z_q}{r_{iq}} - \sum_{jp} \frac{Z_p}{r_{jp}} + \tfrac{1}{2} \sum_{pq} \frac{Z_p Z_q}{r_{pq}} + \tfrac{1}{2} \sum_{if} \frac{1}{r_{if}} \tag{II.1}$$

H_A and H_B are free molecule Hamiltonians for the two molecules, and H_E is the electrostatic interaction. The sums over electrons i and nuclei p, with charges Z_p, belonging to A, and over j and q to B. Taking only the dipole parts of the multipolar expansion of H_E, we have

$$H_E \simeq R^{-3} \mu_i^{(A)} \mu_j^{(B)} \beta_{ij} \tag{II.2}$$

where R is the intermolecular distance, the $\mu_i^{(A)}$ are components of the electric dipole operator for A, and $\beta_{ij} = \delta_{ij} - 3\hat{R}_i\hat{R}_j$, $\hat{R}_i$ being the i-th component of the unit vector along $\underset{\sim}{R}$. The dispersion energy is given by (II.3),

$$\Delta E = - R^{-6} \beta_{ij}\beta_{k\ell} \sum_{n^A, n^B}{}' \frac{\langle 0|\mu_i|n^A\rangle\langle n^A|\mu_k|0\rangle\langle 0|\mu_j|n^B\rangle\langle n^B|\mu_\ell|0\rangle}{E(n^A) + E(n^B)} \quad \text{(II.3)}$$

The energy ΔE has an inverse sixth power dependence on distance, and is scaled by the sum over states of the product of two dipole moment matrix elements from each coupled molecule. The energy denominator is the sum of the excitation energies of the molecular states involved, designated n^A and n^B. It is at once evident that the electric dispersion interaction (II.3) is non-discriminating, inasmuch as replacement of one molecule, say B, by its chiral enantiomer generated by inversion in the origin, changes the signs of the final *two* matrix elements. For each orientation of B, the dispersive coupling to *l*A is the same for *l*B and *d*B. Equality after rotational averaging is obvious. Such detailed considerations are subsumed in the general result that, after isotropic averaging, a difference between *l-l* and *l-d* interaction can appear only if the moment combinations on each centre have a component transforming like a pseudoscalar, namely a quantity invariant to rotation and changing sign under inversion. No binary combination of electric moments has a pseudoscalar component. The lowest order purely electric combination is third-order, where the moment combination $(\underset{\sim}{\mu} \times \underset{\sim}{\mu}).\underset{\sim}{\mu}$ transforms as a pseudoscalar: we may thus expect the third-order pure electric dispersion interaction to exhibit discrimination with distance dependence R^{-9}. It has been discussed by Schipper [8].

II.2 The discriminating dispersion interaction

Inasmuch as the scalar product $\underset{\sim}{\mu}.\underset{\sim}{m}$ of an electric and a magnetic dipole moment transforms as a pseudoscalar, one can foresee that a dispersive interaction in which there is participation by transitions allowed to both electric and magnetic radiation may show discrimination. The appropriate Hamiltonian to replace (II.1) is:

$$H = H_A + H_B + H_E + H_M \quad \text{(II.4)}$$

where

$$H_M = R^{-3} m_i^{(A)} m_j^{(B)} \beta_{ij} \quad \text{(II.5)}$$

and $m_i^{(A)}$ is the i-th component of the magnetic moment operator for molecule A. The dispersion energy with use of (II.4) is given in (II.6)

$$\Delta E = - \sum_{n^A n^B}{}' \frac{\langle 0^A 0^B|H_E + H_M|n^A n^B\rangle\langle n^A n^B|H_E + H_M|0^A 0^B\rangle}{E(n^A) + E(n^B)} \quad \text{(II.6)}$$

This expression includes two terms in addition to the pure electric term (II.3). The pure magnetic term (II.7)

$$\Delta E_{M-M} = - R^{-6} \beta_{ij}\beta_{k\ell} \sum_{n^A n^B}{}' \frac{\langle 0|m_i|n^A\rangle\langle n^A|m_k|0\rangle\langle 0|m_j|n^B\rangle\langle n^B|m_\ell|0\rangle}{E(n^A) + E(n^B)} \quad \text{(II.7)}$$

has the same structure as (II.3), but differs in magnitude. Each appearance of the magnetic moment in place of the electric moment gives a reduction in order of magnitude roughly equal to the fine structure constant $\alpha = 1/137$. The pure magnetic term (II.7) is thus less by 8 or 9 orders than the pure electric (II.3) and, being also non-discriminatory, is of little interest. The second term from (II.6) is the electric-magnetic cross term is given in (II.8),

$$\Delta E_{E-M} = - 2R^{-6} \beta_{ij}\beta_{k\ell} \,\mathrm{Re} \sum_{n^A n^B}{}' \frac{\langle 0|\mu_i|n^A\rangle\langle {}^A|m_k|0\rangle\langle 0|\mu_j|n^B\rangle\langle n^B|m_\ell|0\rangle}{E(n^A) + E(n^B)} \quad \text{(II.8)}$$

With complex wave functions for the states n^A and n^B the expression under the summation sign can have real and imaginary parts, of which only the real part gives a contribution to the energy shift ΔE. Noting that the components of the optical-rotatory tensor $\underset{\approx}{R}$, for the transition $n^A \leftarrow 0$, are

$$(R_{ik})_{0n^A} = \langle 0|\mu_i|n^A\rangle\langle n^A|m_k|0\rangle \quad \text{(II.9)}$$

we have

$$\Delta E_{E-M} = - 2R^{-6} \beta_{ij}\beta_{k\ell} \,\mathrm{Re} \sum_{n^A, n^B}{}' \frac{(R_{ik})_{0n^A} \, (R_{j\ell})_{0n^B}}{E(n^A) + E(n^B)} \quad \text{(II.10)}$$

If we generate the optical isomer of B by inversion in its coordinate origin we change the sign of $(R_{j\ell})_{0n^A}$, and so make a change in the overall sign of ΔE_{E-M}. There is discrimination between *d-d* and *d-l* interactions: moreover the difference persists after isotropic averaging.

Two averages will be discussed, out of many that might be important in various physical situations. In the first it is assumed that the pure electric (non-discriminating) dispersion interaction is strong enough to impose an orientation in which one of the transition dipoles in each molecule included in the sum in (II.10) is aligned along the intermolecular axis. Thus one pair of transitions, one in each molecule, is assumed to dominate both the pure electric dispersive force, and the discrimination, which now becomes (II.11), for real wave functions,

$$\Delta E_{E\text{-}M} = -\ 2R^{-6}\ \beta_{ij}\beta_{k\ell}\ \frac{(R_{ik})_{0n^A}\ (R_{j\ell})_{0n^B}}{E(n^A)\ +\ E(n^B)} \tag{II.11}$$

Now the optical rotatory tensor $\underset{\approx}{R} = \underset{\sim}{\mu}\underset{\sim}{m}$, evaluated for a particular transition, can be split into its symmetric and antisymmetric parts (i.e. with respect to the leading matrix diagonal), according to

$$(\underset{\sim}{\mu}\underset{\sim}{m})^{s} = \frac{1}{2}\ \{(\underset{\sim}{\mu}\underset{\sim}{m})\ +\ (\underset{\sim}{\mu}\underset{\sim}{m})^{T}\}$$

$$(\underset{\sim}{\mu}\underset{\sim}{m})^{a} = \frac{1}{2}\ \{(\underset{\sim}{\mu}\underset{\sim}{m})\ -\ (\underset{\sim}{\mu}\underset{\sim}{m})^{T}\} \tag{II.12}$$

where T indicates the transpose. It is readily shown that the non-diagonal parts of the symmetric tensor vanish on averaging about the electric dipole directions, which lie along the intermolecular axis. The sole contribution is from the diagonal sum, namely the pseudoscalar $R^{\|}$,

$$R^{\|} = \underset{\sim}{\mu}.\underset{\sim}{m} \tag{II.13}$$

The three components of the antisymmetric tensor can be taken to be the components of the vector $\underset{\sim}{\mu} \times \underset{\sim}{m}$ which changes sign under inversion and is therefore polar. The contribution to the average by this part vanishes. Only the parallel part (II.13) survives averaging. For molecules locked with electric dipole transition moments along the intermolecular axis, each being free to rotate independently about it, the averaged discrimination energy is, [10]

$$\Delta E^{1}_{E\text{-}M} = -\ 8R^{-6}\ \frac{R^{\|}(n^A)R^{\|}(n^B)}{E(n^A)\ +\ E(n^B)} \tag{II.14}$$

The minus sign applies when the electric moment directions are parallel and A and B are of like chirality, and becomes plus for antiparallel moments. There is also a sign change for molecules of unlike chirality. Thus the discrimination energy, namely $\Delta E^{1}(d\text{-}d)\ -\ \Delta E^{1}(d\text{-}l)$, is

$$\Delta E^{1}_{Disc} = -\ 16\ R^{-6}\ \frac{R^{\|}(n^A)R^{\|}(n^B)}{E(n^A)\ +\ E(n^B)} \tag{II.15}$$

for molecules with parallel directions of electric dipole moment.

The second average is for random orientations of both molecules. It may be shown to give the discrimination energy (II.16)

$$\Delta E^{1}_{Disc} = -\ (8/3)\ R^{-6}\ \frac{R^{\|}(n^A)R^{\|}(n^B)}{E(n^A)\ +\ E(n^B)} \tag{II.16}$$

as first found by Mavroyannis and Stephen [9]. The absolute signs of (II.15) and (II.16) follow from the fact that, with real wave functions as assumed, the quantity $\mu.m$ is pure imaginary. Thus the overall signs of both right-hand side expressions are positive, and discrimination favours unlike (*d-l*) over like (*d-d*) interactions. The averaged total dispersive interaction is expected to exceed the average dispersive discrimination energy by 3-4 orders of magnitude in representative examples.

III. DISCRIMINATION IN THE RESONANCE INTERACTION

III.1 The discriminating resonance interaction

Two optical antipodes A and B, being chemically identical, have identical energy levels. If we have an AB pair, just as for an AA pair, there is exact resonance between the configurations AB' and A'B, where the prime indicates an excited state, and the well-known electric dipole-dipole resonance interaction causes a splitting into eigenstates (III.1)

$$\left.\begin{array}{l} 2^{-\frac{1}{2}}\,(AB' + A'B) \\ 2^{-\frac{1}{2}}\,(AB' - A'B) \end{array}\right\} \qquad \text{(III.1)}$$

belonging to the total Hamiltonian consisting of free molecule Hamiltonians plus the electric dipolar coupling given in (II.2), namely

$$H_E = R^{-3}\, \mu_i^{(A)}\, \mu_j^{(B)}\, \beta_{ij} \qquad \text{(III.2)}$$

The effect of this term is to split the states (III.1) symmetrically by an amount equal to twice the expectation value of (II.2) evaluated for the transitions A'←A and B'←B. There is also the analogue of (II.2) for the magnetic transition moment of the same transitions in the optically active molecules, giving a result less than (II.2) by four or five orders of magnitude which, in typical cases for a pair of molecules separated by 1 nm, leads to an energy term in the order of 10^2 cm^{-1}.

Both (II.2) and its magnetic analogue are static interactions, with the same form as interactions between permanent electric or magnetic moments. There is no mixed electric-magnetic interaction in this static limit, in contrast to the mixed dispersion interaction (II.8). We shall see however that there is a mixed term when radiative coupling is included between the molecules. This term is dependent on the wavelength of the resonant transition, and at molecular separations of a few nm is comparable in magnitude to the pure magnetic analogue of (II.2). The nature of this interaction is first sketched with the help of classical electrodynamics, as an introduction to the usual quantum electrodynamical picture.

III.2 Radiative electric-magnetic resonance coupling

According to this model, the optically active antipodes A and B act as classical electronic dipole oscillators, characterised by a frequency ω and vectors $\underset{\sim}{\mu}$ and $\underset{\sim}{m}$ equal to the quantal electric and magnetic transition moments for the resonant transition. We suppose that the oscillating electric moment at A, $\underset{\sim}{\mu}e^{-i\omega t}$, is the source of an electromagnetic field which couples to molecule B through its electric moment (giving the pure electric coupling) or through its electric moment (giving the mixed electric-magnetic coupling). To calculate the field at molecule B caused by the source at A we use the retarded Hertz vector $\underset{\sim}{\Pi}$ in the standard way [11], defining

$$\underset{\sim}{\Pi} = \frac{\underset{\sim}{\mu}e^{-i(\omega t-kR)}}{R} \tag{III.3}$$

where R is the separation distance from A to B and k is the wave number, $\omega = ck$. The electric and magnetic fields at B are now found from $\underset{\sim}{\Pi}$,

$$\left.\begin{aligned} \underset{\sim}{E} &= \underset{\sim}{\nabla}\underset{\sim}{\nabla}\cdot\underset{\sim}{\Pi} - \frac{\partial^2\underset{\sim}{\Pi}}{\partial t^2} \\ \underset{\sim}{B} &= \underset{\sim}{\nabla} \times \frac{\partial\underset{\sim}{\Pi}}{\partial t} \end{aligned}\right\} \tag{III.4}$$

giving for the ith component of $\underset{\sim}{E}$

$$E_i^{(B)} = \mu_j^{(A)} e^{-i(\omega t-kR)} \left\{\left(\frac{ik}{R^2} - \frac{1}{R^3}\right) \beta_{ij} + \frac{k^2}{R}\alpha_{ij}\right\} \tag{III.5}$$

where $\beta_{ij} = \delta_{ij} - 3\hat{R}_i\hat{R}_j$ as before, and $\alpha_{ij} = \delta_{ij} - \hat{R}_i\hat{R}_j$. For the magnetic field

$$\underset{\sim}{B}^{(B)} = - i\omega\, e^{-i(\omega t-kR)} \left(- \frac{ik}{R} + \frac{1}{R^2}\right) (\underset{\sim}{\mu}^{(A)} \times \hat{\underset{\sim}{R}}) \tag{III.6}$$

$\hat{R}$ being a unit vector along $\underset{\sim}{R}$. The electric field has a term independent of frequency which is dominant at small values of R (the near zone), and includes a longitudinal component, parallel to $\underset{\sim}{R}$, as well as components transverse to $\underset{\sim}{R}$. The same is true of the term linear in wave number (the 'induction' term). The final term of (III.5) is dominant at large R (the far zone) and is purely transverse, as may readily be seen from the properties of the dyadic α_{ij}. The magnetic field in (III.6) however has no near zone part and is purely transverse in both the far zone and the induction zone, where the field vector varies with distance as R^{-1} and R^{-2} respectively, both terms depending on the frequency. These properties are familiar in the theory of macroscopic electromagnetism, but it is not always realised that molecular phenomena may be discussed in the same way.

The coupling energies of the fields (III.5) and (III.6) with the moments of molecule B are

$$W_{E-E} = - \mu_i^{(A)} \mu_j^{(B)} \left\{ \frac{k^2 \cos kR}{R} \alpha_{ij} - \left(\frac{\cos kR}{R^3} + \frac{k \sin kR}{R^2} \right) \beta_{ij} \right\} \tag{III.7}$$

$$W_{E-M} = \varepsilon_{ijk} \hat{R}_k (\mu_i^{(B)} m_j^{(A)} + \mu_j^{(A)} m_i^{(B)}) \left(- \frac{k^2 \sin kR}{R} + \frac{k \cos kR}{R^2} \right) \tag{III.8}$$

where ε_{ijk} is the Levi-Civita antisymmetric tensor: $\varepsilon_{ijk} = 1$ for cyclic order of suffices, -1 for anticyclic, and 0 for i,j,k not all different.

The electric-magnetic term (III.8) has a leading contribution equal to

$$\frac{\varepsilon_{ijk} \hat{R}_k}{\lambda\!\!\!^{-} R^2} (\mu_i^{(B)} m_j^{(A)} + \mu_j^{(A)} m_i^{(B)}) \tag{III.9}$$

where $\lambda\!\!\!^{-}$ is the reduced wavelength, $\lambda\!\!\!^{-} = \lambda/2\pi = 1/k$, λ being the characteristic wavelength for the molecular transition. At R = 10 nm, λ = 500 nm, this term is less by four orders of magnitude than the electric-electric term (III.7). The term (III.8), but not (III.7), leads to discrimination of this order because replacement of either A or B by its optical isomer changes the sign of one of the two terms in the first bracket. The three resonance coupling terms vanish on random averaging over all orientations.

III.3 The quantum-electrodynamical viewpoint

From the classical point of view, adopted in Sec. III.2, the molecules are treated as emitting and absorbing at a fixed frequency, the coupling between them being through the unquantized electromagnetic field. The same problem has been treated in quantum electrodynamics [10]. Here the starting point is the Hamiltonian

$$H_A + H_B + H_{rad} + H_{int} \tag{III.10}$$

$$H_{rad} = (8\pi)^{-1} \int (E^{\perp 2} + B^2)\, dV \tag{III.11}$$

$$H_{int} = -\underset{\sim}{\mu}^A . \underset{\sim}{E}^{\perp A} - \underset{\sim}{\mu}^B . \underset{\sim}{E}^{\perp B} - \underset{\sim}{m}^A . \underset{\sim}{B}^A - \underset{\sim}{m}^B . \underset{\sim}{B}^B \tag{III.12}$$

where H_A and H_B are free molecule Hamiltonians, $\underset{\sim}{E}^{\perp}$ denotes the transverse electric field, and the integral in (III.11) is taken over the space occupied by the system.

According to (III.10) the radiation field is a part of the unperturbed system, whereas in the classical picture it had the emitting molecule as its source. The interaction operator H_{int} splits the states (III.1) in the second order of perturbation theory. The energy shifts for the + and - states include contributions from pure electric terms W_{E-E} as before, pure magnetic W_{M-M}, and magnetic-electric W_{E-M}. The latter is

$$W_{E-M}(R) = \pm \sum_{m,\lambda} \left\{ \frac{\langle A'B;0|\mu_i^A E_i^{\perp A}|AB;k\rangle\langle AB;k|m_j^B B_j^B|AB';0\rangle}{E - \hbar\omega} - \frac{\langle A'B;0|m_j^B B_j^B|A'B';k\rangle\langle A'B';k|\mu_i^A E_i^{\perp A}|AB';0\rangle}{E + \hbar\omega} + \text{terms with } A \leftrightarrow B \right\} \qquad \text{(III.13)}$$

The states appearing in the matrix elements are composite states of molecules and field, the latter being denoted by the wave vector $\underset{\sim}{k}$ and the polarization index λ, which is summed over in (III.13). The frequency ω is no longer the fixed oscillator frequency used in Sec. III.2, but a variable $\omega = ck$ summed over the range of the radiation frequencies. The matrix elements in (III.13) correspond to the time-ordered graphs (a) and (b) in Fig. 3.

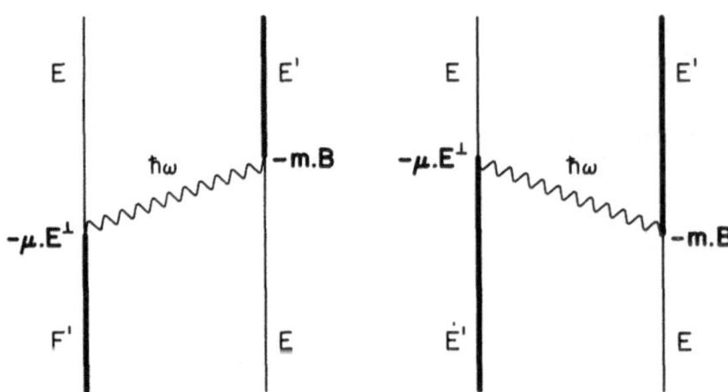

Figure 3. Time ordered graphs for the two terms shown explicitly in expression (III.13)

After a little manipulation we find results for the energy shifts identical with those found heuristically in the last section, including the electric-magnetic cross-term (III.9) for the processes in the graphs of Fig. 3. The results now however are based on a quantal treatment of both molecules and field.

IV. ELECTROSTATIC INTERACTIONS

IV.1 Choice of multipole origin

If the external electric field of a distribution of charges in a neutral molecule or ion is expanded in point multipole moments at an origin, the coefficients in the expansion take different values according to the origin choice. The interaction energy of two charge distributions, at distances at which the distributions do not overlap, is invariant to origin choice so long as the expansion of the electric potential (or field) is extended indefinitely. If it is cut off, the energy is origin dependent. In molecules where the origin can be placed at a centre of symmetry the potential expansion includes only spherical harmonics of even order, whereas for other choices harmonics of all orders are in general required. In chiral molecules, lacking a centre of symmetry, the most convenient origins are points invariant to as many as possible of the molecular symmetry operations. The only multipole moments allowed to appear in the potential expansions at such points are those invariant to the symmetry operations, and simplifications may be possible. In chiral molecules of point groups $\underset{\sim}{D}_2$ and $\underset{\sim}{D}_3$ the origin is taken at the intersection of the principal rotation axis and the perpendicular twofold axes: with that choice the potential expansions contain no dipole components and there are other restrictions. In groups $\underset{\sim}{C}_n$ the origin is taken, for similar reasons, at a point on the rotation axis. In asymmetric systems (group $\underset{\sim}{C}_1$) there is no symmetry basis for the choice of origin, and no limitation on the multipole components that may appear in the expansion. It is rather useful in $\underset{\sim}{C}_1$ and $\underset{\sim}{C}_n$ systems to place the origin at the centre of mass, as it is already placed by symmetry in the $\underset{\sim}{D}_2$ and $\underset{\sim}{D}_3$ systems. Interactions between molecules rotating freely may then be averaged over orientation angles keeping the separations fixed, as measured between centres of mass, inasmuch as the rotation of each molecule is about this point as centre. In problems not involving the rotational average, including interactions in fixed relative orientations, other criteria for the origin choice have been used, such as making one or more multipole components vanish.

IV.2 Chiral combinations of multipoles

The electric field produced by the charge distribution of a chiral molecule can be expanded in a series of point multipolar fields located at the centre of mass. It is important to find what properties the multipole moments must possess to simulate the chiral character of the charge distribution [12]. Writing the electric multipole moments $Q(n\ell)$ in terms of spherical harmonics

$$Q(n\ell) = \sum_i \{e_i r_i^n\} P_n(\cos\theta_i) e^{i\ell\phi_i} \qquad \text{(IV.1)}$$

where the sum is over the charges in the distribution, we may readily find the transformation properties. For molecules with symmetries

$\underset{\sim}{C}_n$ and $\underset{\sim}{D}_n$, the polar axis is placed along the symmetry axis, and the transformation under improper rotation by $2\pi/p$ is given in (IV.2),

$$iC_p \; Q(n\ell) = (-1)^n \; e^{2\pi i\ell/p} \; Q(n\ell) \qquad \text{(IV.2)}$$

Reflection in a plane containing the symmetry axis and making an angle ξ with the molecular x axis induces the transformation (IV.3)

$$iC_2'(\xi) \; Q(n\ell) = e^{2i\xi} \; Q(n-\ell) \qquad \text{(IV.3)}$$

The charge distribution can be represented by the real multipole moment components (IV.4) all of which have at least one plane of symmetry. A minimum

$$\left.\begin{aligned} Q_n^{\ell+} &= \frac{1}{2}\,\{Q(n\ell) + Q(n,-\ell)\} \\ Q_n^{\ell-} &= \frac{1}{2i}\,\{Q(n\ell) - Q(n,-\ell)\} \end{aligned}\right\} \qquad \text{(IV.4)}$$

requirement for chirality in the potential source is a combination of two multipole moments. It is readily confirmed that for a chiral molecule of $\underset{\sim}{C}_n$ symmetry the minimum, lowest-order, chiral combination of two moments is of orders n and n+1. In $\underset{\sim}{C}_3$, for example, the combinations $\{Q_3^{3\pm}, Q_4^{3\pm}\}$ are chiral, the upper and lower sign pairs being taken together. In groups $\underset{\sim}{D}_n$ a similar requirement holds, and $\{Q_n^{n+}, Q_{n+1}^{n+}\}$ describes a chiral combination of moments [12].

In asymmetric molecules, group $\underset{\sim}{C}_1$, we are free to choose the body axis system arbitrarily, with origin at the centre of mass. The simplest chiral combination is most easily visualised as a quadrupole component, say Q_2^{2+} combined with a dipole oriented to have no reflection plane in common with it. Thus in terms of components in a body-fixed axis system we require two dipole components, say Q_1^0 and Q_1^{+1}, combined with Q_2^{2+}.

At short distances of separation the multipole expansion method is often of little use. Convergence is slow, and the values of the multipole moment components are not known. The electrostatic energy could be calculated better with the help of the point charge model, where the external molecular field is represented as that of point charges placed on the atoms, with magnitudes estimated empirically from bond moments or found from approximate molecular wave functions. An alternative is to represent the molecular field by two dipoles, placed at different points in the molecule, and oriented to give chiral character [1].

IV.3 Interaction between chiral multipole combinations

As is well known the interaction energy of a pair of electric multipoles, one on each of two centres held at fixed separation, vanishes on averaging randomly over all orientations. There is no electrostatic discrimination between such freely rotating systems. Evidently also if we have an assembly of fixed chiral molecules without orientational correlation there can be no electrostatic discrimination. This might be taken to be the instantaneous situation in solution. Equally however the intermolecular or interionic forces may be strong enough to impose preferred orientations, leading to discrimination energies which can be in the order of several hundred J mol^{-1} under realistic assumptions. For example it has been shown [12] that two chiral dipole-quadrupole combinations, in *l* and *d* forms, show substantial discrimination between *l*-*l* and *l*-*d* pairs. The dipole moments are assumed large enough to impose a head-to-tail collinear arrangement. The systems are free to rotate about the axis. Then, with quadrupole components of 1 em^{-20} and separation between systems of 0.5 nm, the most stable *d*-*d* configuration lies 300 J mol^{-1} below the most stable *d*-*l* configuration. Such magnitudes would be of the right order to account for many observed discriminations.

V. DISCRIMINATION IN TRANSIENT CHIRAL SPECIES

V.1 Discrimination involving achiral molecules in a chiral medium

In the experimental examples of discrimination given in Sec. I the evidence is in most cases in terms of thermodynamic quantities, such as heats of solution and mixing, changes in redox potentials, and free energies of optical inversion ('enantiomerisation'). A different system altogether is now described in which the strength of discriminating interactions may be connected with the circular dichroism of one partner of an interacting pair of molecules [2,13].

The key to the discussion is the Jahn-Teller effect on the degenerate excited electronic states of certain achiral molecules. The equilibrium structures in these excited states may be chiral, and the enantiomers are acted upon differently by the intermolecular forces coupling them to any other chiral species present, such as the molecules of a chiral solvent. The discrimination is then expected to be observed through the appearance of circular dichroism.

The Jahn-Teller theorem states that a molecule in a degenerate electronic state is unstable to displacements in one or more symmetry coordinates. The active displacements are those which carry the molecule from a group of higher symmetry into one of lower symmetry lacking degenerate representations, or perhaps one in which the bases for degenerate representations are different. Thus, to take a familiar example (Fig. 4) an e_{2g} displacement taking the benzene molecule from the D_{6h} structure into a D_{2h} structure is Jahn-Teller active in a degenerate electronic state of species E_{1u}, splitting this degenerate state into two non-degenerate states belonging to different representations of the lower symmetry group. In the present context we seek a Jahn-Teller displacement into a chiral structure, and the lower symmetry group must be one of the chiral groups $\underset{\sim}{C}_n$ or $\underset{\sim}{D}_n$. We find that for n=2 the

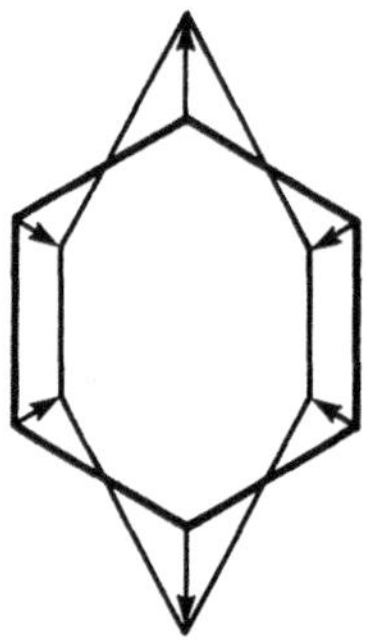

Figure 4. Effect of an e_{2g} symmetry coordinate displacement on a $\underset{\sim}{D}_{6h}$ object, leading to residual $\underset{\sim}{D}_{2h}$ symmetry

simplest achiral 'parent' groups, admitting degenerate states, are $\underset{\sim}{D}_{2d}$ and $\underset{\sim}{S}_4$, both of which have fourfold improper symmetry axes. Examples of $\underset{\sim}{D}_{2d}$ molecules are allene (Fig. 5), *spiro*nonatetraene [of which the displaced form is shown in Fig. 1(c)] and *spiro*-bifluorene.

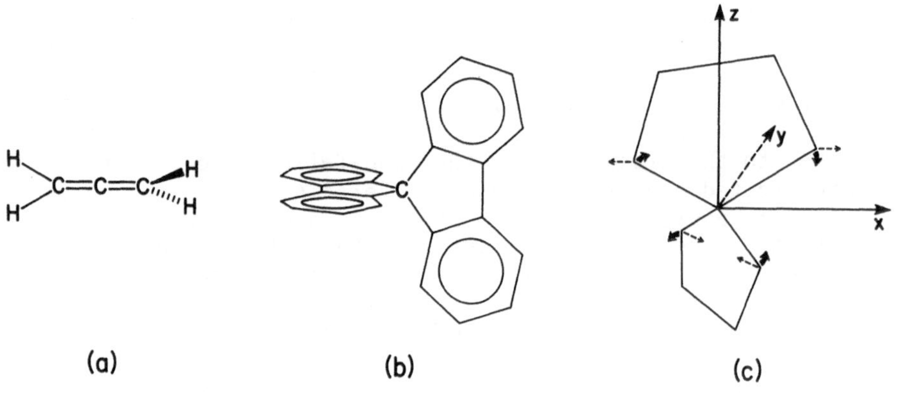

Figure 5. (a) Allene (b) *spiro*-bifluorene (c) Typical Jahn-Teller active displacements in *spiro*nonatetraene. b_1 shown by full arrows and b_2 by dotted arrows

A nuclear displacement (species Γ_v) is Jahn-Teller active in a degenerate electronic state of symmetry species Γ_e if the direct product $\Gamma_e \times \Gamma_v \times \Gamma_e$ contains the totally symmetrical representation. In $\underset{\sim}{D}_2$

we thus find that displacements of species b_1 and b_2 are active. Examples of such displacements are shown in Fig. 5(c). The character table for D_{2d} with the z axis along the improper fourfold axis is given in Table 1. It shows that if there is a nuclear displacement of symmetry b_1, the resulting structure retains the symmetry operation E, C_2^x, C_2^y and C_2^z and thus belongs to D_2, the representations of which,

Table 1. Character Table for D_{2d}

D_2 rep.	D_{2d} rep.	E	C_2	$2S_4$	$2C_2'$	$2\sigma_d$
A	A	1	1	1	1	1
B	A	1	1	1	-1	-1
A	B	1	1	-1	1	-1
B	B	1	1	-1	-1	1
B_2, B_3	E	2	-2	0	0	0

in their association with those of the parent D_{2d}, are shown in the left-hand column of Table 1. The degenerate representation of D_{2d} is split into B_2 and B_3 representations of D_2. If on the other hand the original structure is displaced along a b_2 coordinate of D_{2d}, the structure retains the symmetry elements E, C_2 and the two reflections σ_d, thus belonging to the achiral C_{2v}, which is of no interest here.

In allene the displacements corresponding to those of Fig. 5(c) are the torsion about the S_4 axis, and the antisymmetric scissor motions of the terminal CH_2 groups. Following excitation to the degenerate electronic excited state the molecules are unstable in symmetry D_{2d} and respond by going into vibrational motions under potentials of which the minima are displaced from the ground state minimum along b_1 or b_2 symmetry coordinates or some combination of them. This lasts for the lifetime of the excited state, which may be as long as the radiative lifetime of 10^{-9}-10^{-8}s but will in most cases be shortened by collisions. During the period the average nuclear configuration, in the case of b_1 instability, will be that for a (chiral) D_2 molecule. In the absence of forces favouring one or other of the enantiomers, the two forms formed by right- and left-hand torsions about the z axis will occur with equal probability. If chirally discriminating forces act one enantiomer is favoured.

V.2 Spectroscopic implications

Following a straightforward adaptation of the ideas of Child [14] and Hougen [15] we write the leading terms in the vibronic Hamiltonian for a D_{2d} molecule, including the effect of one displacement coordinate q_1 belonging to species b_1 and one q_2 belonging to b_2. The Hamiltonian (V.1)

$$H = H_{e\ell} - \frac{\hbar^2}{2m_1}\frac{\partial^2}{\partial q_1^2} - \frac{\hbar^2}{2m_2}\frac{\partial^2}{\partial q_2^2} + \frac{1}{2}k_1q_1^2 + \frac{1}{2}k_1q_1^2 + L_1q_1 + L_2q_2 \quad \text{(V.1)}$$

includes $H_{e\ell}$, the electronic Hamiltonian for the $\underset{\sim}{D}_2$ structure. m_1 and m_2 are effective masses for the two normal vibrations, k_1 and k_2 the force constants and L_1 and L_2 the linear coupling coefficients characteristic of the Jahn-Teller effect. To simplify the discussion we suppose that $L_2=0$; this is a reasonable assumption, for example, in allene, where the effect of the b_2 CH_2 scissor motion on the lower electronic levels must be very much smaller than that of the b_1 axial torsion. Now for any $L_1 \neq 0$ we find there are two new equilibrium positions displaced by equal distances $\pm L_1/k_1$ along q_1. The electronic states, derived from E or $\underset{\sim}{D}_{2d}$ are B_2 and B_3 of the new group $\underset{\sim}{D}_2$, and since the vibrations in the coordinate q_1 are both of species a_1 in $\underset{\sim}{D}_2$ the vibronic species are B_2 and B_3. Fig. 6(a) shows the situation in an achiral medium.

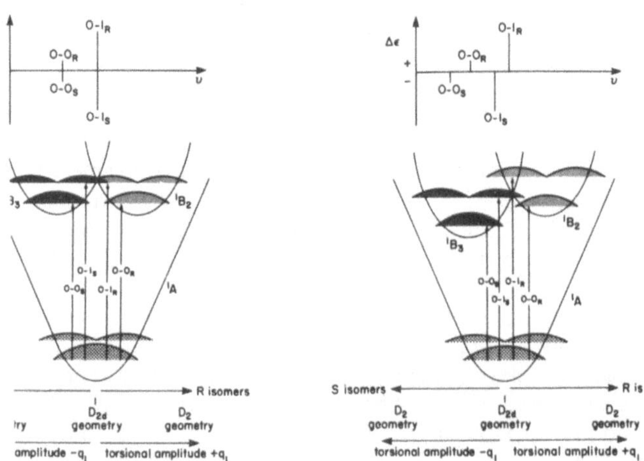

Figure 6. The variation of electronic energies with the torsional coordinate q, for a D_{2d} molecule (a) in an achiral environment (b) in a chiral environment. The spectral transitions are labelled with the R and S enantiomer labels. Vibrational probability distributions are indicated. Vibronic contributions to the CD spectrum cancel exactly in (a) but not in (b). Taken, with permission, from Chem. Phys. Letters, 41, (1976), 225.

Transitions from the ground state zero-point level end at levels of the two electronic excited states with strength proportional to the squared Franck-Condon overlap integrals over the ground state and excited state vibrational wave functions. There is no circular dichroism, inasmuch as the left- and right-circular absorptions are at precisely equal frequencies as shown in the schematic spectrum. Fig. 6(b) applies when one enantiomer of another chiral species is present, which stabilises one of the two distorted structures relatively. Spectral transitions to the more stable upper state appear at the longwave onset of the spectrum and can show CD over a wavelength range determined by the energy gap between the $^1B_2$ and $^1B_3$ levels.

V.3 Some magnitudes

One can see that the observed strength and frequency width of the circular dichroism depends on at least four quantities, if one disregards the effects of temperature on level populations and various sources of line broadening. The quantities are the magnitude of the Jahn-Teller displacement, the discrimination energy in the interactions with an added chiral solute or with a chiral solvent, the intrinsic CD in the transitions shown in Figs. 6(a) and 6(b), and the Franck-Condon factor for the transition at the onset of the spectrum, which determines what fraction of the total CD for the transition can appear in a chosen vibronic transition. The Franck-Condon factor depends upon the size of the Jahn-Teller displacement.

For the Jahn-Teller effect we next draw attention to *ab initio* calculations of the total electronic energy of *spiro*nonatetraene [13] as a function of torsion angle about the *spiro* carbon centre. The energy minimum is calculated to be at an angular displacement from the $\tilde{D}_{2d}$ structure of about ±2.5°, the minima being about 110 cm^{-1} (1.31 kJ mol^{-1}) below undistorted $\tilde{D}_{2d}$. The Franck-Condon factor for such an angular displacement is in the order of 10^{-1}, probably in the range 0.3-0.7.

The strength of CD is proportional to the rotatory strength for the transition n←0

$$R^{no} = \mathrm{Im}\ \langle 0|\tilde{\mu}|n\rangle\cdot\langle n|\tilde{m}|0\rangle \qquad (V.2)$$

where $\tilde{\mu}$ and $\tilde{m}$ are the operators for electric and magnetic dipole moments. In the present case, the symmetry-determined zero circular dichroism of $\tilde{D}_{2d}$ *spiro*nonatetraene in the transition from the ground state to the degenerate E state can be attributed to a cancellation of the rotatory strengths separately belonging to transitions to the two real degenerate components E_α←A and E_β←A, according to

$$\sum_{i=\alpha,\beta} \mathrm{Im}\ \langle 0|\tilde{\mu}|E_i\rangle\langle E_i|\tilde{m}|0\rangle = 0 \qquad (V.3)$$

The CD observed at the onset of a spectrum of type 6(b) is thus proportional to the rotatory strength of one of the transitions, in the spectral region where there is no cancellation with the second transition. The value found in [13] is about 0.5 Debye Bohr magnetons, which is typical of values for a large number of chiral molecules.

The last of the quantities, the discrimination energy between the two D_2 structures in the presence of added chiral molecules, is much more difficult to estimate in a convincing way. As already discussed the largest discriminations are those that have their origin in the superior packing of one form or the other with adjacent chiral molecules. In our case, where the chiral structures are transient, with lifetimes less than 10^{-9}s and perhaps as short as a few picoseconds, the local relaxation leading to optimum local packing will usually have a *longer* lifetime, and the differential packing possibilities will not in practice be realised. The discriminations that are expected to be observed are therefore those for which the diastereomeric pairs are at equal separation distances. The probable sources of discrimination are thus the electrostatic and dispersive interactions.

Even in the D_2 structure a molecule cannot possess a static electric dipole moment. Electrostatic discrimination is expected to be less than the values earlier quoted for that reason and also because of the reduction of electrostatic terms by orientational averaging (Sec. IV). Dispersive terms may however contribute, but it is as already seen unlikely to exceed 10 cm^{-1}. The electrostatic terms would probably not exceed 100-200 cm^{-1} even under the most favourable conditions. Thus one would deduce that the splitting between $^1B_2$ and $^1B_3$ in Fig. 6(b) is toward the lower end of the range 1-200 cm^{-1}; this is the spectral range in which CD should be capable of observation in molecules of the *spiro*nonatetraene type. Measurement of the spectral range of the CD which directly gives the discrimination energy under these conditions of little or no contribution by differential packing (short range) terms, would be of exceptional value in setting the scale of chiral discriminations in general.

REFERENCES

[1] Craig, D.P. and Mellor, D.P.: 1976, Topics in Current Chemistry 63, 1.
[2] Mason, S.F.: 1976, Ann. Rep. Chem. Soc. 73, 53.
[3] Craig, D.P. and Stiles, P.J.: 1976, Chem. Phys. Letters 41, 225.
[4] Dwyer, F.P. and Davies, N.R.: 1954, Trans. Faraday Soc. 50, 24.
[5] Barnes, G.T., Backhouse, J.R., Dwyer, F.P. and Gyarfas, E.C.: 1956, Proc. Roy. Soc. N.S.W. 89, 151.
[6] Chion, B., Lajzerowicz, J., Collet, A. and Jacques; J.: 1976, Acta Cryst. B32, 339.
[7] Perucand, M.C., Cauceill, J. and Jacques, J.: 1974, Bull. Soc. Chim. France 1011.
[8] Schipper, P.E.: 1978, Chem. Phys. 28, 357.
[9] Mavroyannis, C. and Stephen, M.J.: 1962, Mol. Phys. 5, 629.
[10] Craig, D.P., Power, E.A. and Thirunamachandran, T.: 1971, Proc. Roy. Soc. A322, 165.

[11] Stratton, J.A.: 1941, *Electromagnetic Theory*, McGraw Hill, New York and London.
[12] Craig, D.P. and Schipper, P.E.: 1975, Proc. Roy. Soc. A342, 19.
[13] Craig, D.P., Stiles, P.J., Palmieri, P. and Zauli, C.: 1979, J. Chem. Soc. Far. II 75, 97.
[14] Child, M.S.: 1960, Mol. Phys. 3, 601.
[15] Hougen, J.T.: 1964, J. Mol. Spectry. 13, 149.

SYMPOSIUM IV. NEW FIELDS OF MOLECULAR SPECTROSCOPY

Chairman: J. A. Pople

Department of Chemistry, Carnegie-Mellon University,
Pittsburgh, Pennsylvania, U.S.A.

NEW FIELDS IN MOLECULAR SPECTROSCOPY

John A. Pople
Carnegie-Mellon University, Pittsburgh, PA 15213 U.S.A.

INTRODUCTORY REMARKS

As an introduction to the lectures that follow in this Symposium, some general comments on overall directions in which spectroscopy has recently moved may be useful. These new developments increase the power of experimental techniques in exploring all aspects of molecular structure. At the same time, they provide exciting challenges to theoretical chemists in interpreting and rationalizing new data that is emerging.

Classical spectroscopy was the science of emission and absorption of single photons, leading to a change of molecular state. The main objective in most early studies was the use of an energy balance to determine the energy difference between the molecular states involved. In addition, simple expressions for transition probabilities for low intensity radiation were developed (dipole transition theory) and used to obtain well-known dipole selection rules. The latter have proved valuable in the identification and classification of molecular states.

We may trace several ways in which spectroscopy has expanded from its initial base. All of these will be touched on in the following lectures and should stimulate some interactive discussion.

1. Spectroscopy and external perturbations. The stationary-state energy levels of a molecule can be altered by various external perturbations which are under the control of the experimenter. Applied strong electric and magnetic fields are the most important of these, leading to the well-developed topics of Stark and Zeeman spectroscopy. Most forms of magnetic resonance spectroscopy fall into this category, where transitions are observed between levels separated by an external magnetic field. The theoretical challenge here is to understand the nature of the perturbation introduced by the external field. Although a lot of work has been done, theoretical treatments

K. Fukui and B. Pullman (eds.), Horizons of Quantum Chemistry, 147–149.

of such effects as diamagnetic polarization and the related NMR chemical shifts are still very inadequate.

2. Spectroscopy and inelastic scattering. Transitions between stationary states of molecules can be induced in ways other than by absorption or emission of single photons. This leads to the integration of classical spectroscopy with a variety of other inelastic scattering properties. Single-photon absorption spectroscopy corresponds to

$$h\nu + \text{molecule} \rightarrow \text{molecule}$$

Other processes are:

$$h\nu + \text{molecule} \rightarrow \text{molecule}^* + h\nu' \quad \text{(Raman spectroscopy)}$$

$$e + \text{molecule} \rightarrow \text{molecule}^* + e' \quad \text{(Inelastic electron scattering)}$$

Some involve ionization:

$$h\nu + \text{molecule} \rightarrow \text{molecule}^+ + e \quad \text{(Photoelectron spectroscopy)}$$

$$e + \text{molecule} \rightarrow \text{molecule}^+ + e' + e'' \quad \text{(e, 2e spectroscopy)}$$

In all of these processes, the energies of incoming and outgoing particles (photons or electrons) may be measured so that an energy balance can be set up.

3. Spectroscopy and momentum changes. An important aspect of spectroscopy and the general inelastic scattering studies is that studies can be made of the direction of emitted particles (usually electrons) as well as their energies. Thus it is possible to set up a momentum balance as well as an energy balance. The study of intensity as a function of angle relative to the incoming beam provides more information about the nature of the transition than could be obtained by the simple dipole selection rules. In ionization spectroscopy, these momentum studies can lead to detailed properties of the molecular orbital from which the electron is removed. Again, this is an area to which quantum chemists should be able to contribute.

4. Spectroscopy and non-linear effects. An aspect of spectroscopy which is of increasing importance is the possibility of studying the effects of strong radiation fields so that non-linear effects become significant. In quantum terms, these correspond to multiphoton processes. In addition to straightforward saturation effects, there is the possibility of 'double-resonance' type of experiments in which one transition is studied while another (involving one of the same energy levels) is simultaneously subjected to radiation. This type of study originated in magnetic resonance spectroscopy where it is relatively easy to achieve radiation fields strong enough to

observe nonlinear processes. However, the advent of various forms of laser spectroscopy has led to other types of double-resonance spectroscopy, some of which will be mentioned by the following speakers.

MOLECULAR SPECTROSCOPY USING INFRARED LASERS; A STUDY OF RADIATIVE AND COLLISIONAL PROCESSES

Takeshi Oka
Herzberg Institute of Astrophysics
National Research Council of Canada
Ottawa, Ontario, Canada, K1A 0R6

It is nearly 20 years since the first lasers were successfully operated. The first decade after the discovery was spent mostly in the development of a great variety of lasers and in studies of new phenomena, but their enormous potential as radiation sources for spectroscopy has lately become increasingly clear as a result of many new experiments. For example, in the four symposia of the series on laser spectroscopy that started in 1973 [1,2,3,4], many novel and ingenious experiments have been published. The two key quality factors of spectroscopy, that is, the sensitivity and the resolution, have been increased by many orders of magnitude. Parallel to these activities mainly by physicists, attempts at applying this new radiation source to systematic molecular spectroscopy have been made mainly by chemists. Over the last several years in particular, there has been an exponential growth in the activity in this field. The present paper is concerned with a very small portion of this field directly related to my own work. Emphasis will be placed on various new phenomena with their theoretical implications rather than on details of individual molecules.

1. THE TWO ELEMENTARY PROCESSES

The two elementary processes in gaseous molecules to be discussed are shown in Figure 1. One is the radiative process (Figure 1a) in which a molecule interacts with the radiation field and changes its quantum state. The other is the collisional process (Figure 1b) in which a molecule comes close during its motion to another molecule and changes its quantum state as a result of intermolecular interaction. The discussion will be focussed on the rotational state of molecules because the new phenomena

K. Fukui and B. Pullman (eds.), Horizons of Quantum Chemistry, 151–167.

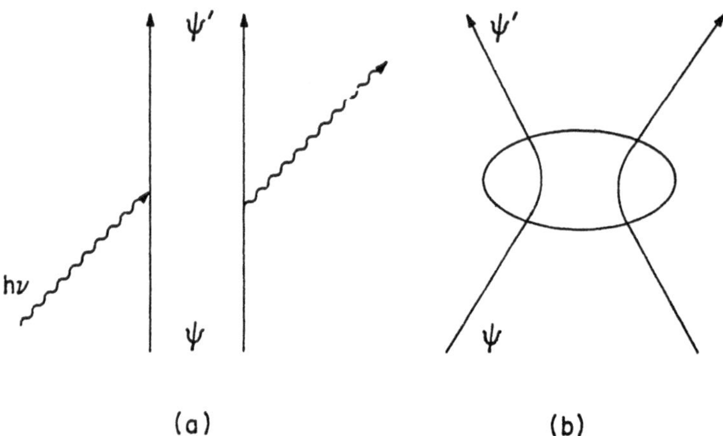

Fig.1. Two elementary processes: (a) the radiative process and (b) the collisional process.

described in this paper appear most clearly there.

It is normally said in textbooks that the radiative processes are governed by rigorous selection rules and the collisional processes occur in a random fashion. That is, if we specify the initial state Ψ of the molecule, the final state Ψ' is one of only a few possible states allowed by the selection rules for radiative transitions whereas it could be any of hundreds of other states for collisional transitions. The radiative process is well behaved but the collisional process is chaotic.

What I would like to show by experimental evidence is that if we look more carefully, the radiative process is not that well behaved, that is, we can observe "forbidden" transitions that are not obeying the normal selection rules; also the collisional processes are not that random, that is, we can observe many subtleties of the processes. The reason behind these observations is seen if we consider the interaction energy of these two processes. The energy for the radiative interaction is μE where μ is the molecular transition dipole moment and E is the electric field of the radiation. We consider it in units of h, that is the Rabi frequency μE/h. The energy for the collisional interaction is the intermolecular potential energy V.

The radiative process is well behaved in normal spectroscopy because the interaction energy is very small.

We can treat it by the usual time dependent perturbation method which has a good convergence. However if we use a strong radiation field such as that of microwave or laser radiation, the applied radiation field E is much higher. Such a radiation field can easily be 1~10^3 V/cm as opposed to the typical value of 10^{-3} V/cm for normal spectroscopy and in certain cases, such as multiphoton dissociation by the pulsed high power CO_2 laser, can be of the order of 10^7 V/cm. This increase of field changes the situation in two ways. (a) The convergence of the time dependent perturbation gets worse and higher order effects such as two-photon and multiphoton processes occur with a high probability. This breaks the normal selection rules $|\Delta J| \leq 1$ of the total angular momentum and $+ \leftrightarrow -$ of the parity. (b) The intense electric field reveals various effects of subtle intramolecular interactions which are normally neglected. Such interactions produce dipole moments which are orders of magnitude smaller than those normally considered. By applying a high radiation field, we can make the Rabi frequency appreciable even for such a small dipole moment. A few examples of these cases will be presented in Section 2.

The collisional process on the other hand is often thought to be random because the intermolecular potential $V \sim kT$ is large compared to the rotational energy spacings and cannot be treated by perturbation procedure. However this applies only to strong head-on collisions. There are much weaker collisions for which the interaction energy is orders of magnitude smaller than kT and yet molecules apply sufficient torque to each other to change the rotational state. If we can monitor these weak collisions, we can detect subtle effects as we see for radiative processes. Some examples will be presented in Section 3.

To summarize this section, I think that there is not that much difference between the two processes. If a transition is allowed by collisional processes it is also allowed by radiative processes and vice versa. The purpose of this paper is to give experimental evidence for this theme.

2. RADIATIVE PROCESSES

Let us start from the study of radiative processes in which the normal selection rules are violated. First let me remind you of the basic logic involved in this discussion, shown in Figure 2.

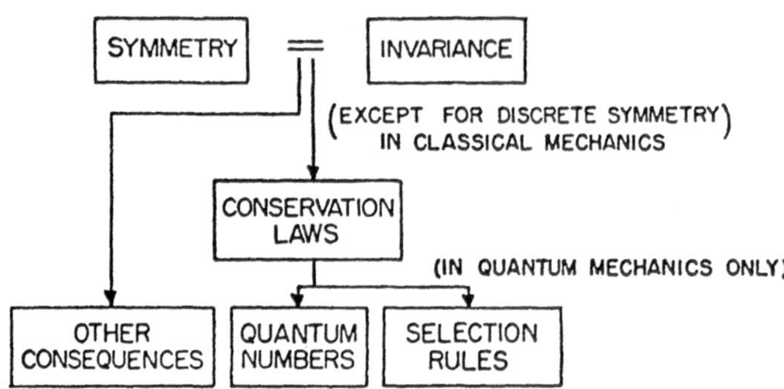

Fig.2. Relation between symmetry and selection rules [5]

Suppose we consider NH_3 as an example. The rotation-vibration levels of this molecule can be specified by giving three quantum numbers, J for the total angular momentum, k for its projection along the molecular symmetry axis, and the parity. The normal radiative selection rules are $\Delta J = 0, \pm 1$, $\Delta k = 0$, and $+ \leftrightarrow -$. The quantum number J and parity arises from the symmetry of space, that is, the isotropy and the reflection symmetry, while the quantum number k arises from the cylindrical symmetry of the equilibrium inertia tensor of the molecule. The selection rules for the former are violated when the symmetry of space is broken by, for example, a high radiation field, and that for the latter when the geometrical symmetry of the molecule is broken by, for example, centrifugal distortion. While the former multiphoton process is treated as a higher order time-dependent perturbation, the latter (forbidden transition) is treated as a time-independent perturbation.

The two-photon processes due to externally applied radiation(s) were first observed in the radiofrequency region and later extended to microwave-microwave [6], microwave-optical [7,8] and optical-optical [9] radiation(s). The last process is particularly significant for spectroscopy because by the use of two counter-propagating radiations, Doppler broadening can be eliminated. Multiphoton processes were observed by using more powerful pulsed CO_2 lasers [10]. In such an experiment often a power density of the order of 10^{11} Watt/cm^2 is used. This power density translates to 10^7 V/cm of radiation field and to the Rabi frequency of 1.5×10^{12} Hz or 50 cm^{-1} for a dipole moment of 0.3 Debye. It will be much higher for higher vibrational states. For such a high value of Rabi frequency, the selection rules for J are randomized. This fact is neglected so far in the discussion of multiphoton dissociation and isotope separation [11].

The breakdown of geometrical symmetry of a molecule and the resulting violation of selection rules for the related quantum numbers such as k results from intramolecular interaction of higher order. Instead of the dipole selection rule $\Delta k=0$, we have the more relaxed rule $\Delta k=3n$. This latter rule results from the permutation symmetry of the three equivalent protons and is harder to violate because the mixing interactions have to change the nuclear spin state and are very small.

The $\Delta k=3n$ transitions were first observed in collisional processes [12] but later studied also for radiative processes [13,14]. Figure 3 shows radiative relaxation of NH_3 through spontaneous emission of such transitions.

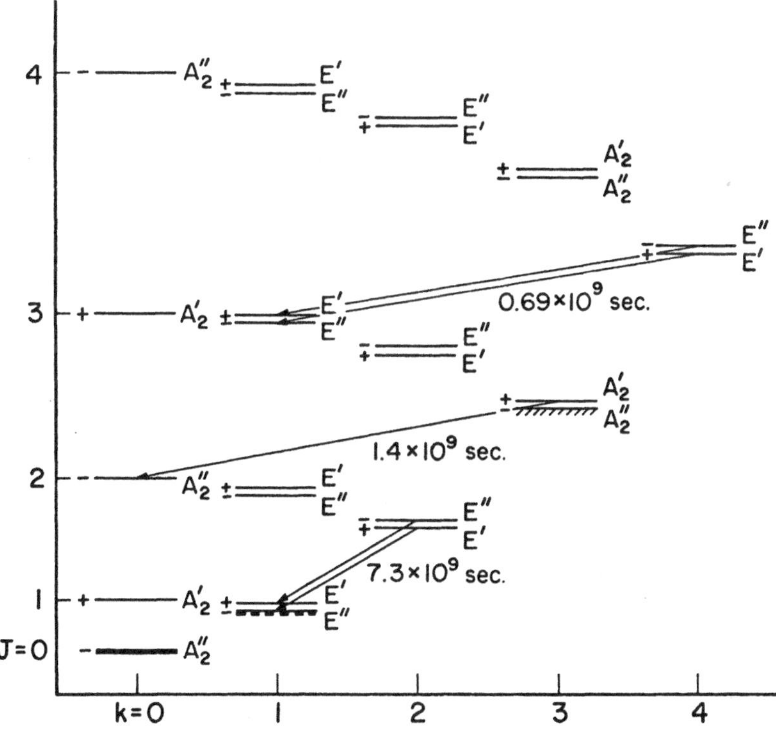

Fig.3. Rotation-inversion energy levels of NH_3 and spontaneous emission time for the $\Delta k=\pm 3$ transitions from the metastable rotational levels.

Normal $\Delta k=0$ transitions keep the molecules within one column of the same k but the $\Delta k=3$ transitions can mix molecules in different columns. The latter transitions are of course much slower than the former and take many years to occur but in the astronomical scale of time are sufficiently fast to be seriously considered. The laboratory observations of such transitions were done for PH_3, PD_3 and AsH_3 [15,16] and POF_3 [17].

A particularly interesting case of the breakdown of geometrical symmetry is observed for tetrahedral molecules [14,18,19]. The centrifugal distortion of the molecules produces a small dipole moment of the order of 10^{-4} J(J+1) Debye (Figure 4).

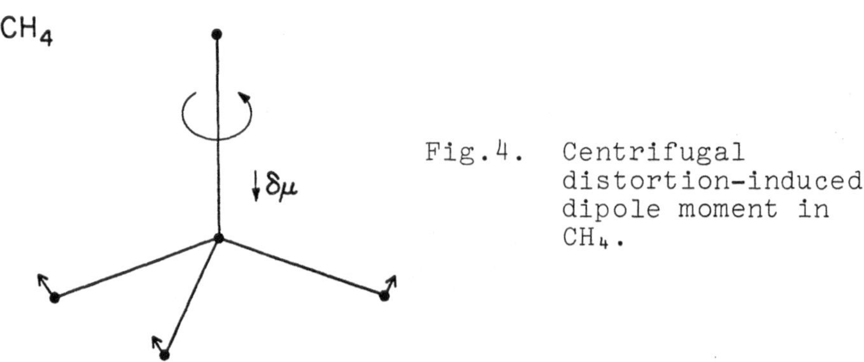

Fig.4. Centrifugal distortion-induced dipole moment in CH_4.

The Stark shift of rotational levels due to such a dipole moment in CH_4 is indeed measurable [20]. Thus a tetrahedral molecule is polar if it is not in the J=0 level. Observation of pure rotational transitions in the radiofrequency region through such a dipole moment was carried out using the technique of laser-radiofrequency double resonance. The outline of the apparatus is shown in Figure 5.

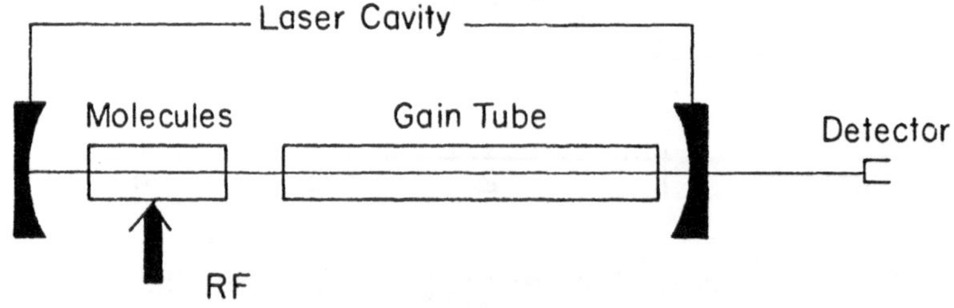

Fig.5. Basic experimental set-up for radiofrequency spectroscopy inside a laser cavity.

We place a sample of gas at a low pressure (∿ 5 mTorr) in a laser cavity and apply strong radiofrequency (or microwave) radiation. The frequency of the rf radiation is swept while the laser output power is monitored. When the rf frequency comes to resonance, the bulk characteristics of the gas change as the laser load changes and the resonance is detected as a sharp variation of the laser output power. In this operation the laser is providing infrared radiation for the double resonance and at the same time working as part of the detection system. This method is extremely sensitive for detecting molecular resonances in the radiofrequency region [see Ref. 8 and 21 for more details].

The radiofrequency pure rotational spectra in tetrahedral molecules have been detected by using this sensitive method in CH_4 [22], SiH_4 [23], and GeH_4 [24]. Fig. 6 shows an example of GeH_4. The coincidence between the 10 P(26) line of the CO_2 laser and the Q(11) line of the ν_2 band of GeH_4 is used [a]. The observed signal with the Stark shift is shown in [b]. We can see that under the

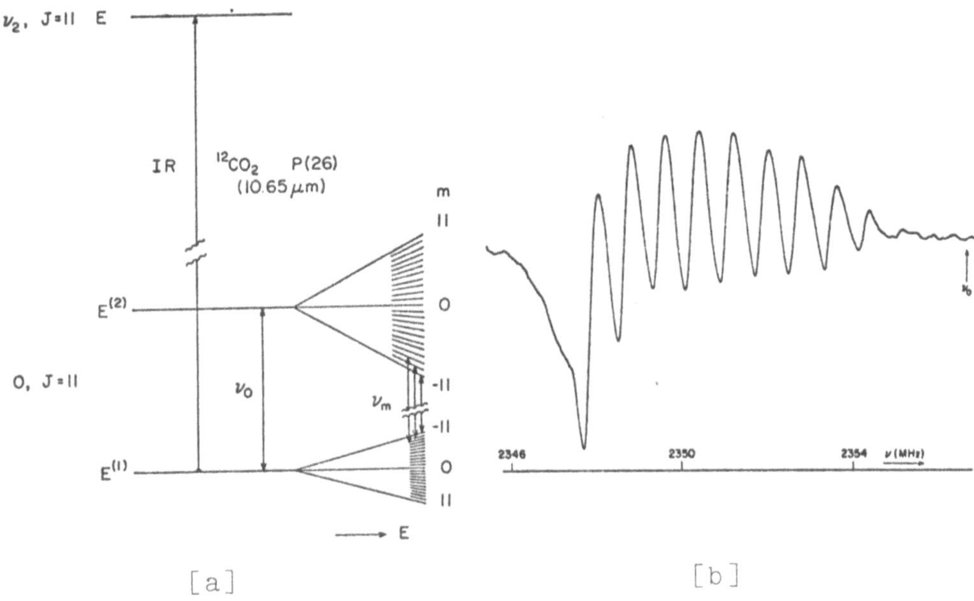

Fig.6. [a] The energy level system for an infrared-microwave double resonance in GeH_4.
[b] Observed J=11 $E^{(2)} \leftarrow E^{(1)}$ forbidden rotational transitions of GeH_4 with the first order Stark pattern [25].

high sensitivity detection method, the tetrahedral molecule behaves just like a polar molecule.

When we apply this extremely sensitive method to normal polar molecules, we see a great many lines. Figure 7 shows such an example of CF_3I.

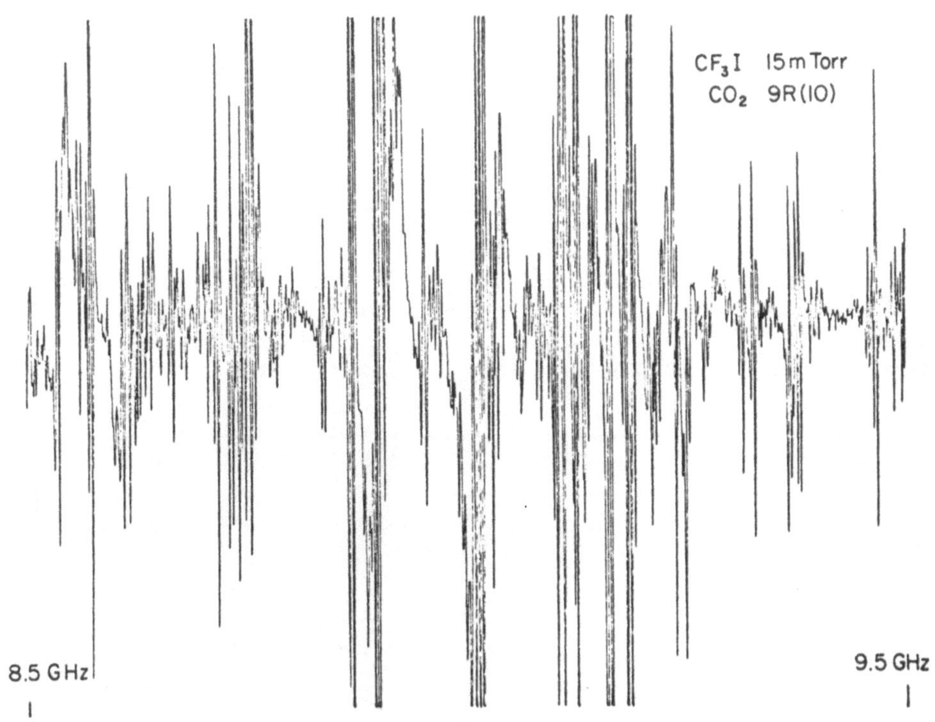

Fig.7. Example of dense spectra for heavy polar molecules [26].

Here the 9 R(10) CO_2 laser line and microwave radiation between 8.5 and 9.5 GHz were used. This is the infrared region where the isotope separation of ^{13}C was reported by the method of multiphoton dissociation [27]. The enormous number of lines are all genuine resonances, not noise, and we obtain this type of spectrum for the other rf region and other laser lines. Although we have not attempted to assign these lines, the number of lines is far greater than that expected from normal selection rules. We believe this relaxation of selection rules also plays an important part in the discussion of multiphoton dissociation especially for higher vibrational states.

In summary ample evidence has been accumulated for the existence of rotational transitions which are not expected from the normal selection rules. In fact there is no pair of levels in a molecule which is not connected by radiative processes. Even transitions between different species (which result from the most stubborn symmetry of nuclear permutations) have been observed for near degenerate cases [28,29,30].

3. COLLISIONAL PROCESSES

We now turn our attention to the second molecular process - the collisional process. The first evidence which had shown that such processes do not occur in a random fashion but obey some sort of "selection rules" was a four-level microwave double resonance experiment [31]. In this experiment two microwave radiations ν_p and ν_s were used. The first radiation ν_p "pumps" a molecular transition and thus introduces non-Boltzmann populational distributions in the two levels associated with the transition. This anomalous population is then transferred to other levels by collisional processes. The second radiation ν_s traces this transfer by monitoring the variation in intensities of other transitions. By using various combinations of ν_p and ν_s, the existence of selection rules has been established for many molecules [32].

With the advent of the laser the four-level double resonance experiment has been extended to infrared-microwave double resonance [33] and to infrared-infrared double resonance [34]. These methods have since been applied to many molecules and confirmed and extended the results obtained by microwave double resonance. One new aspect of the infrared-infrared double resonance experiment which was lacking in the microwave double resonance experiment is its capability to monitor the molecular velocity along the line of sight because of the large Doppler width. By using this capability, we can not only monitor the change of internal molecular state but also the variation of molecular velocity due to collisional processes. In the following we will concentrate on this aspect of the infrared double resonance experiments.

Figure 8 shows what happens in these experiments.

The strong infrared radiation ν_p pumps molecules from level 1 to level 3. Because of the large Doppler width of the infrared spectrum ($\nu v/c \sim 30$ MHz), ν_p pumps only those

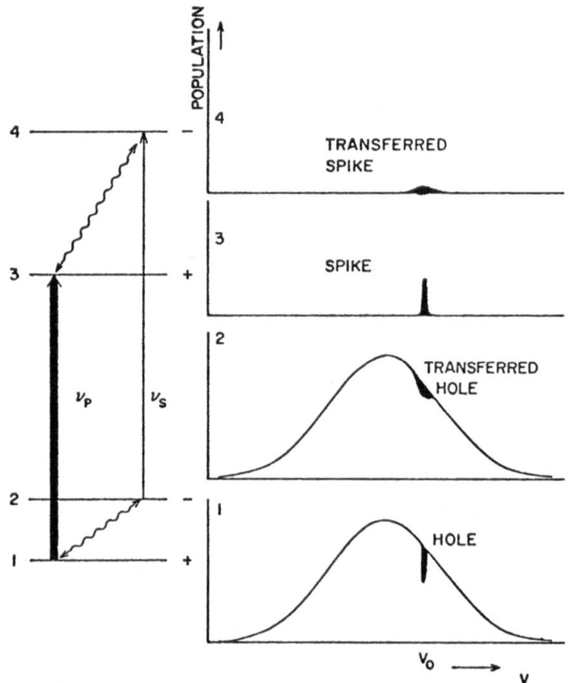

Fig.8. The four-level infrared-infrared double resonance experiments [34].

molecules with velocity v which satisfies $(E_3-E_1)/h = \nu_p(1+v/c)$ thus creating a "hole" in the Maxwellian velocity profile of molecules in level 1 and a "spike" in level 3. These popularion anomalies are then transferred to levels 2 and 4 by collisional processes and are monitored by the second radiation ν_s. If molecules change their velocities very much when they change their quantum states through collisional processes, the hole and spike will be smeared out. The experiment using the NH_3 molecule showed rather clearly that they are not smeared out [34]. This means that the NH_3 molecules change their inversion states without appreciably changing their velocity; the velocity memory is kept.

We have seen evidence of this velocity conservation in two other experiments quite unexpectedly. The second one was the collision-induced centre dip in Stark spectroscopy [35,36,37]. An example is shown in Figure 9.

The two intense signals in Figure 9 are normal Lamb dips in laser Stark spectroscopy corresponding to molecules with v=0. The small signal at the centre, however, is caused by collision-induced transfer of the hole and the

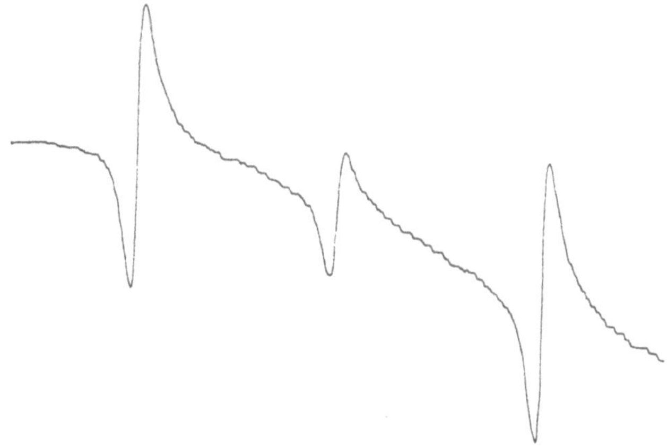

Fig.9. Collision-induced centre dips in the Stark Lamb dip spectroscopy in the $\nu_3Q(1,1)$ transitions of CH_3F [38].

spike for molecules with certain non-zero velocity. The sharpness of such collision-induced centre dip is additional evidence that molecules do not appreciably change their velocities when they change their rotational levels. In this case the rotational transition is between different Stark components, that is, molecules are re-orienting ($\Delta M \neq 0$) with respect to space without changing their speed and internal orientation of rotation ($\Delta J = \Delta K = 0$). The intensity of the collision-induced centre dip relative to those of normal Lamb dips gives a measure of the efficiency of the reorientation of molecules. This was shown to be very high in CH_3F, more than 50% of the total rotational transitions [37].

The third example of velocity conserving collisional processes was observed in "pure" nuclear quadrupole resonances using the infrared-radiofrequency double resonance apparatus shown in Figure 5. Figure 10 gives an example in CH_3I. The very high sensitivity of this detection method reveals many collision-induced resonances as seen in Figure 10b. These are due to velocity conserving collision-induced transitions between $\Delta J \neq 0$ levels of CH_3I which cause the transfer of holes and spikes to many rotational levels from those pumped directly by the laser radiation. The relative intensities of these collision-induced signals give information on relative probabilities of various collisional processes.

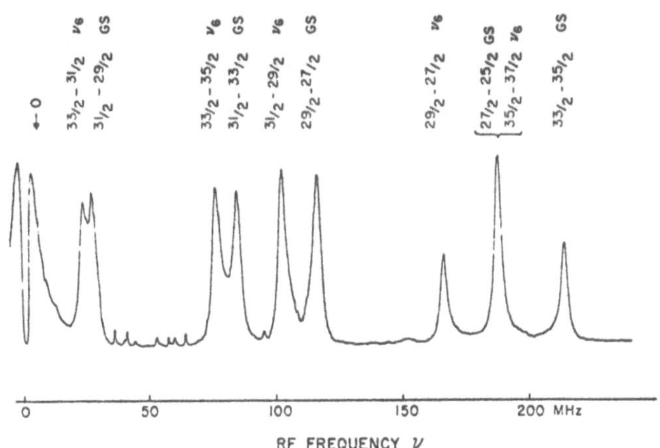

(a)

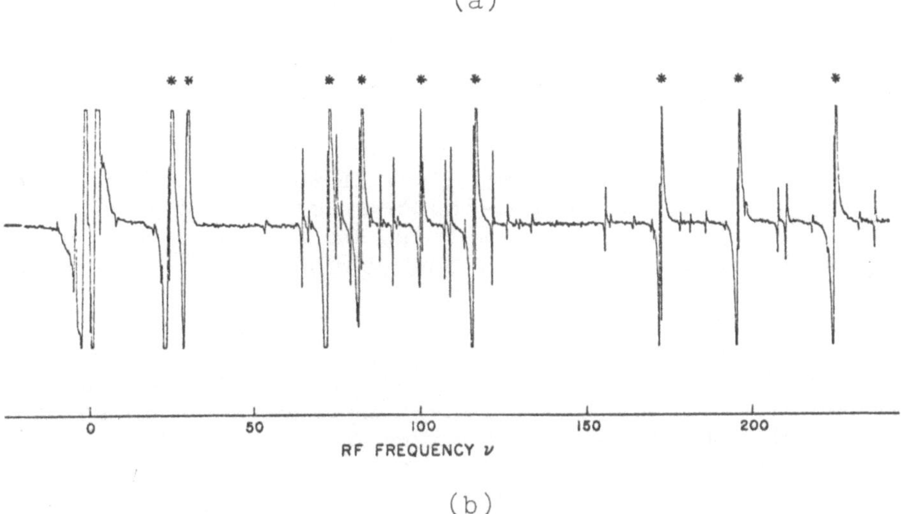

(b)

Fig.10. "Pure" nuclear quadrupole resonances in CH_3I gas (∿ 10 mTorr). The coincidence between the $\nu_6{}^rR(15,5)$ transition of CH_3I and the 10 P(32) CO_2 laser line is used. (a) Basic pattern of two sets of quintets. (b) The same spectrum taken with the high sensitivity detection mode. Many collision-induced resonances are observed in addition to the original pattern (marked with *) [39].

So far I have given experimental evidence for the velocity conserving collision in which a molecule changes its internal state without changing its velocity appreciably. Now I would like to discuss the opposite of this process, that is, a collisional process in which a molecule changes its velocity without changing its internal state. These two extreme processes are shown schematically in Figure 11.

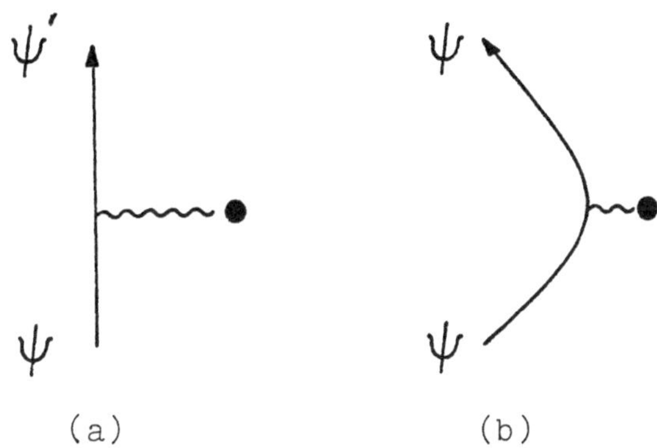

Fig.11. Two extreme cases of the collisional process.
(a) Velocity conserving inelastic collision;
(b) Velocity changing elastic collision.

The second case shown in Figure 11b is observed by the use of laser spectroscopy as the Dicke narrowing [40] of infrared spectra. When many of this type of collision occur consecutively, the molecule changes its velocity many times while keeping its quantum state and thus interacting with radiation of the same wavelength. The Doppler shift of the infrared spectrum is then determined by the average velocity $\langle v \rangle$ over the many molecular paths rather than the instantaneous velocity v for each path between collisions. Since the former is smaller than the latter, the Doppler width of the spectrum is reduced. This effective averaging of velocity occurs when the mean free path of molecules is comparable or smaller than the wavelength of radiation. Thus the Dicke narrowing is easiest to observe for microwave spectra of atoms [41].

For molecules, the observation of Dicke narrowing is more difficult because molecules do not stay in a given quantum state as stubbornly as atoms do; the transitions between rotational levels occur very easily. The Dicke narrowing for molecules was first observed in the Raman spectrum of hydrogen at high pressure [42,43]. The Doppler limited infrared laser spectroscopy gives convenient means to observe this phenomenon fairly easily for a large number of molecules because of the longer wavelength of radiation [44]. Figure 12 shows an example of the narrowing of the HCl $v = 1 \leftarrow 0$ R(13) line narrowed by Ar atoms.

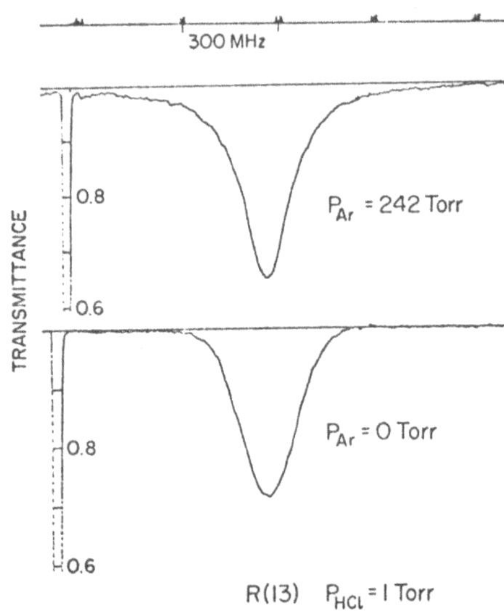

Fig.12. Dicke narrowing observed for the HCl-Ar mixture. The lower trace shows the Gaussian Doppler broadened line shape for pure HCl (1 Torr) while the upper trace shows the Lorentzian Dicke narrowed line shape for the HCl:Ar mixture (1 Torr:242 Torr) [45].

A difference frequency laser system [46] has been used to observe the narrowing. The analysis of the observed linewidths is pictorially shown in Figure 13. The linewidth is decomposed into pressure broadening and Doppler broadening

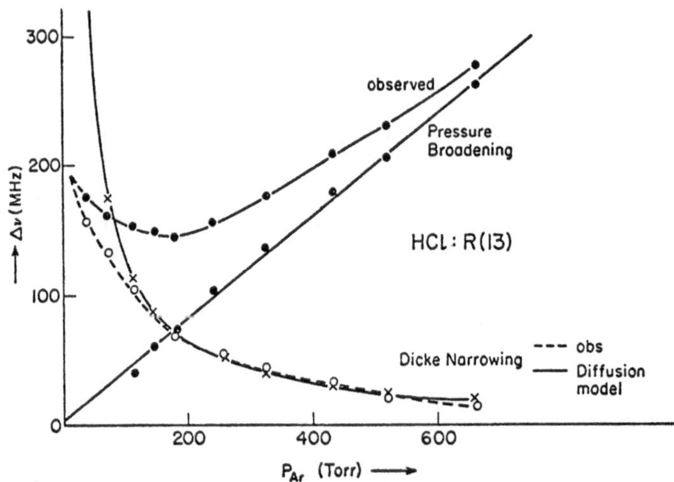

Fig.13. Analysis of the linewidth. The observed linewidths are decomposed into pressure broadening width and the Dicke narrowed Doppler width.

which increases and decreases with pressure, respectively.

I have shown examples of two extreme cases in the above discussion, because extreme cases are always easier to observe and simpler to analyze. However most of the actual intermolecular collisions will correspond to intermediate cases where the change of internal state is accompanied by variation of molecular velocities. The capability of lasers to monitor molecular velocities will enable us to study such processes. We can perform four-level double resonance and scattering experiments simultaneously without using molecular beams. I feel most enthusiastic to carry out these experiments in the next several years.

I wish to thank J.K.G. Watson for reading this paper.

REFERENCES*

[1] "Laser Spectroscopy" (Proc. 1st Laser Spectroscopy Conf. Vail, Colorado) edited by R.G. Brewer and A. Mooradian (Plenum, New York, 1974).

[2] "Laser Spectroscopy" (Proc. 2nd Laser Spectroscopy Conf., Megève) edited by S. Haroche, J.C. Pébay-Peyroula, T.W. Hänsch and S. Harris (Springer-Verlag, Berlin Heidelberg New York, 1975).

[3] "Laser Spectroscopy III" (Proc. 3rd Laser Spectroscopy Conf., Wyoming) edited by J.L. Hall and J.L. Carlsten (Springer-Verlag, Berlin Heidelberg New York, 1977).

[4] "Laser Spectroscopy IV" (Proc. 4th Laser Spectroscopy Conf., Rottach-Egern) edited by H. Walther and K.W. Rothe (Springer-Verlag, Berlin Heidelberg New York, 1979).

[5] C.N. Yang, 1978, Nobel Lecture, Science 124, pp. 565-569.

[6] T. Oka and T. Shimizu, 1970, Phys. Rev. A 2, pp. 587-593.

[7] T. Oka and T. Shimizu, 1971, Appl. Phys. Lett. 19, pp. 88-90.

[8] See for a summary, T. Oka, "Frontiers in Laser Spectroscopy, II" (Proc. Summer School of Theoretical Physics, Les Houches) edited by R. Balian, S. Haroche and S. Liberman (North-Holland Pub. Co., 1977) pp. 529-569.

* This list of references is not meant to be exhaustive.

[9] F. Biraben, B. Cognac and G. Grynberg, 1974, Phys. Rev. Lett. 32, pp. 643-645.
M.D. Levenson and N. Bloembergen, 1974, Phys. Rev. Lett. 32, pp. 645-648.

[10] N.R. Isenor and M.C. Richardson, 1971, Appl. Phys. Lett. 18, pp. 224-228.

[11] V.S. Letokhov, "Frontiers in Laser Spectroscopy, II" (Proc. Summer School of Theoretical Physics, Les Houches), edited by R. Balian, S. Haroche and S. Liberman (North-Holland Pub. Co., 1977 (pp. 771-868.

[12] T. Oka, 1968, J. Chem. Phys. 49, pp. 3135-3145.

[13] T. Oka, F.O. Shimizu, T. Shimizu and J.K.G. Watson, 1971, Ap.J. Lett. 165, pp. L15-L19.

[14] J.K.G. Watson, 1971, M. Mol. Spectrosc. 40, pp. 536-544.

[15] F.Y. Chu and T. Oka, 1974, J. Chem. Phys. 60, pp. 4612-4618.

[16] D.A. Helms and W. Gordy, 1977, J. Mol. Spectrosc. 66, pp. 206-218; 69, pp. 473-481.

[17] R.H. Kagann, I. Ozier and M.C.L. Gerry, 1977, Chem. Phys. Lett. 47, pp. 572-574; 1978, J. Mol. Spectrosc. 71, pp. 281-298.

[18] K. Fox, 1971, Phys. Rev. Lett. 27, pp. 233-236.

[19] See for a summary, T. Oka, "Molecular Spectroscopy: Modern Research, Vol. II" (Academic Press, New York, 1976), pp. 229-253.

[20] I. Ozier, 1971, Phys. Rev. Lett. 27, pp. 1329-1332.

[21] E. Arimondo, P. Glorieux, and T. Oka, 1978, Phys. Rev. A 17, pp. 1375-1393.

[22] R.F. Curl, Jr., T. Oka, and D.S. Smith, 1973, J. Mol. Spectrosc. 46, pp. 518-520.
R.F. Curl, Jr., 1973, J. Mol. Spectrosc. 48, pp. 165-173.

[23] W.A. Kreiner and T. Oka, 1975, Can. J. Phys. 53, pp, 2000-2006.

[24] W.A. Kreiner, U. Andresen, and T. Oka, 1977, J. Chem. Phys. 66, pp. 4662-4665.

[25] W.A. Kreiner, B.J. Orr, U. Andresen and T. Oka, 1977, Phys. Rev. A 15, pp. 2298-2304.

[26] E. Arimondo, V. Lukashenko and T. Oka, unpublished.

[27] S. Bittenson and P.L. Houston, 1977, J. Chem. Phys. 67, pp. 4819-4824.

[28] I. Ozier, P.N. Yi, A. Khosla, and N.F. Ramsey, 1970, Phys. Rev. Lett. 24, pp. 642-646.

[29] I. Ozier and W.L. Meerts, 1978, Phys. Rev. Lett. 40, pp. 226-229.

[30] Ch. J. Bordé, M. Ouhayoun, A. Van Lerberghe, C. Salomon, S. Avrillier, C.D. Cantrell and J. Bordé, "Laser Spectroscopy IV", edited by H. Walther and K.W. Rothe (Springer-Verlag, Berlin Heidelberg New York, 1979).

[31] T. Oka, 1966, J. Chem. Phys. 49, pp. 754-755.

[32] See for a summary, T. Oka, 1973, Adv. At. Mol. Phys. 9, pp. 127-206.

[33] T. Shimizu and T. Oka, 1970, J. Chem. Phys. 53, pp. 2536-2537.
T. Shimizu and T. Oka, 1970, Phys. Rev. A 2, pp. 1177-1181.

[34] S.M. Freund, J.W.C. Johns, A.R.W. McKellar and T. Oka, 1973, J. Chem. Phys. 59, pp. 3445-3453.

[35] M. Ouhayoun and C. Bordé, 1972, C.R. Acad. Sci. B272, pp. 411-414.

[36] R.G. Brewer, R.L. Shoemaker and S. Stenholm, 1974, Phys. Rev. Lett. 33, pp. 63-66.

[37] J.W.C. Johns, A.R.W. McKellar, T. Oka, and M. Römheld, 1975, J. Chem. Phys. 62, pp. 1488-1496.

[38] U. Andresen and T. Oka, unpublished.

[39] E. Arimondo, P. Glorieux and T. Oka, unpublished.

[40] R.H. Dicke, 1953, Phys. Rev. 89, pp. 472-473.

[41] J.P. Wittke and R.H. Dicke, 1956, Phys. Rev. 103, pp. 620-631.

[42] V.G. Cooper, A.D. May, E. Hara and H.F.P. Knapp, 1968, Can. J. Phys. 46, pp. 2019-2023.

[43] J.R. Murray and A. Javan, 1969, J. Mol. Spectrosc. 29, pp. 502-5-4.

[44] R.S. Eng , A.R. Calawa, T.C. Harman, P.L. Kelly and A. Javan, 1972, Appl. Phys. Lett. 21, pp. 303-305.

[45] D. Ramachandra Rao and T. Oka, 1979, unpublished.

[46] A.S. Pine, 1974, J. Opt. Soc. Amer. 64, pp. 1683-1690.

MOLECULES AS SOURCES

P.W. Atkins
Physical Chemistry Laboratory,
University of Oxford,
England.

ABSTRACT. Schwinger's theory of sources is applied to problems of molecular scattering, spectroscopy, and interaction. It is shown that the two fundamental universal principles of causality and space-time uniformity can be used to construct expressions for the scattering amplitudes of basic molecular processes. The main part of the paper is a simplified development of the ideas and techniques of source theory, the aim being to formulate them in a way suitable for non-relativistic molecular processes. It is shown in detail how to use causality and uniformity to construct the amplitudes that appear in the S-matrix formulation of optical birefringence and dispersion interactions.

1. INTRODUCTION

The electromagnetic properties of molecules can be discussed very readily and systematically using S-matrix theory (1-3). Essentially one is interested in the evolution of some prepared initial state $|-\infty\rangle$ into some detected final state $|\infty\rangle$ as a result of an interaction. For instance, in optical activity $|-\infty\rangle$ may be the direct product of a molecular state and an n-photon plane-wave polarization state; then the detected part of $|\infty\rangle$ will be the direct product of the original molecular state, but the photon state in the forward direction will be a superposition of the original state with a component of the orthogonal polarization state. The two states are connected by the operator $\mathbb{S}$: $|\infty\rangle = \mathbb{S}|-\infty\rangle$, and so all the information about the interactions leading to the distortion of the combined initial state, and all the information about the outcome of detection experiments, is embedded in $\mathbb{S}$. This formalism is useful for the systematic development of higher-order interactions (nonlinear effects) and for the discussion of new kinds of electromagnetic examinations of molecules, such as photon coincidence scattering experiments (3). In applications one is more concerned with the component of $\mathbb{S}$ that leads to transitions out of the initial state; this means that one concentrates on calculating $\mathbb{T} = \mathbb{S}-1$; the $\mathbb{T}$-matrix can very readily and systematically be expressed diagrammatically (1-3) in a way that is helpful in organizing orders of

K. Fukui and B. Pullman (eds.), Horizons of Quantum Chemistry, 169–190.

perturbation interactions and which shows the structure of the underlying molecule-field interaction processes.

In this paper the aim is to examine the basis of the type of calculation referred to above. In particular, the intention is to show how much can be deduced about the electromagnetic properties of molecules on the basis of two very fundamental and primitive ideas: causality and space-time uniformity. We shall see, in fact, that these two principles are sufficient to lead to expressions that are recognizable as $\mathbb{S}$-matrix and $\mathbb{T}$-matrix (or $\mathbb{R}$-matrix) elements. Since we have already established that the $\mathbb{S}$ and $\mathbb{T}$ matrices can be used to discuss both dispersive and absorptive processes (as well as elastic and inelastic processes), this constitutes a complete programme.

The discussion (4,5) follows the theory introduced by Schwinger (6-9) in a series of books and articles. The general philosophy is as follows. A particle is to be defined by the collisions that create it, and the role of the other particles taking part in the collision can be reduced to their idealization as a source of the required properties. Thus, in optical activity the optically active molecule is to be regarded merely as a source of the orthogonal polarization photon state, and in a Raman experiment, the scattering target molecule is the source of the frequency-shifted emergent photon. The detection step is also viewed in a similar fashion: there the particles (the term particles including photons) hand on their properties. In the detection process the particles assume the role of sources, and the detector is a sink. The causality principle enters by recognizing that the sinks and sources are localized and ordered in space and time, and any composite process has to be expressed as an ordered sequence of events. The principle of space-time uniformity enters through the requirement that there be nothing to distinguish one component of the source from another except the region of space-time it occupies. On the basis of these general fundamental principles about the nature of the world, Schwinger has been able to construct a theory of interactions which, he claims, avoids the mathematical ambiguities and physical remoteness of operator field theory and the mathematical attitude and speculative philosophy of the supposedly more physical S-matrix theory. This paper does not claim such strong advantages when source theory, as the general approach is called, is used in the discussion of molecular electromagnetic interactions; but it has been concluded elsewhere (4) that source theory does add to our understanding of the electromagnetic interactions of molecules, and provides the theoretician with an elegant but difficult technique for exploring their basic structure.

2. GENERAL FORMALISM

In this section we examine how the two fundamental principles of causality and uniformity are used constructively. The development is less general than Schwinger's original, for it is the intention to confine attention to molecular spectroscopic, dispersive, and scattering processes.

The basic constructive element of the theory is the amplitude with which the source operates. (The term source will be used to denote either an emissive source or an absorptive sink: they are essentially the same, but respond differently, just as a radio receiver and transmitter are essentially the same but respond differently.) The vacuum before the operation of a source is denoted $|0_{-}\rangle$; the source S operates to produce a 1-particle state with 4-momentum P, $|1_P\rangle$, and does so with an amplitude we shall denote $\langle 1_P(S)0_{-}\rangle$. When the source acts as a sink it annihilates a 1-particle state, and produces the vacuum $|0_{+}\rangle$ with a success expressed by the amplitude $\langle 0_{+}(S)1_P\rangle$. The $\pm$ subscripts are causal labels pertaining to the operation of the specified source. The source itself may also generate particles dressed more comprehensively than with only 4-momentum, in which case the amplitudes have to be labelled more explicitly; for example $\langle 1_{P\pi}(S)0_{-}\rangle$ is the source amplitude for the generation of a single polarized (π) particle, and $\langle 1_{PJM}(S)0_{-}\rangle$ is the amplitude for a molecule with rotational energy. The sources themselves generate characteristic particles. For instance, a vector source generates a vector particle (e.g. a photon), a real scalar source generates a neutral spin-0 particle, a complex scalar source generates a charged spin-0 particle, and a J-rank tensor source generates a molecule in its J-rotational state. Spinor sources generates spinor particles, such as electrons.

For a variety of applications it is convenient to separate out from the source amplitudes various factors that disappear on integration, and to introduce Fourier transforms in order to take account of the extension of the sources in space-time. We introduce the following definitions:

$$\langle 1_P(S)0_{-}\rangle = iS_P = iS(P)\sqrt{d\omega_P} \tag{2.1}$$

$$\langle 0_{+}(S)1_P\rangle = iS_P^{*} = iS^{*}(P)\sqrt{d\omega_P} \tag{2.2}$$

$$d\omega_P = d\vec{p}/(2\pi)^3 2P^0 \tag{2.3}$$

$$S(P) = \int dR\,S(R)\exp(-iP.R) \tag{2.4}$$

where R is the 4-vector $(\vec{r},t)$, $P = (\vec{p},E)$, and $A\cdot B = \vec{a}\cdot\vec{b} - a^0b^0$.

The basic elements, the source amplitudes, can be combined according to rules derived on the basis of causality and uniformity. Take the action of a source $S^{(-)}$, which generates particles, followed by the action of a source (sink) $S^{(+)}$ which destroys them, and consider the amplitude for the persistence of the vacuum after both sources have operated. This amplitude is the amplitude of a composite source, $\langle 0_{+}(S^{(+)}+S^{(-)})0_{-}\rangle$, and it can be expressed as the superposition of the amplitudes for the causal operation of the individual sources. In the case of weak real scalar sources,

$$<0_+(S^{(+)}+S^{(-)})0_-> = <0_+(S^{(+)})0_-><0_+(S^{(-)})0_->$$

$$+ \sum_P <0_+(S^{(+)})1_P><1_P(S^{(-)})0_-> + \ldots.$$

$$= <0_+(S^{(+)})0_-><0_+(S^{(-)})0_-> + \sum_P iS_P^{(+)*} iS_P^{(-)} + \ldots.$$

$$= <0_+(S^{(+)})0_-><0_+(S^{(-)})0_-> +$$

$$\int d\omega_P dR^{(+)} dR^{(-)} iS^{(+)}(R^{(+)}) \exp\{iP\cdot(R^{(+)}-R^{(-)})\} iS^{(-)}(R^{(-)}) + \ldots \tag{2.5}$$

Now invoke the second principle, the uniformity in space-time. The implication of this principle is that in the overall description of a process, there should be nothing to distinguish one component of the source from another, apart from the region of space-time each one occupies. It follows that the vacuum persistence amplitude following the operation of the joint source $S = S^{(+)} + S^{(-)}$ must be of the form

$$<0_+(S)0_-> = 1 + \tfrac{1}{2}i\int dRdR'S(R)\Delta_+(R-R')S(R') . \tag{2.6}$$

On comparison of this expression with the preceding one it follows that

$$\Delta_+(R-R') = i\int d\omega_P \exp\{iP\cdot(R-R')\} \qquad \text{for } t > t' \tag{2.7a}$$

$$= i\int d\omega_P \exp\{iP\cdot(R'-R)\} \qquad \text{for } t < t' \tag{2.7b}$$

or

$$\Delta_+(R-R') = \left(\frac{1}{2\pi}\right)^4 \int dP \left(\frac{\exp\{iP\cdot(R-R')\}}{\{P^2+ m^2-i\varepsilon\}}\right) . \tag{2.7c}$$

An immediate consequence is that $\Delta_+(R-R')$ is the causal propagator for a Klein-Gordon equation, since it satisfies

$$(-\partial^2+ m^2)\Delta_+(R-R') = \delta(R-R') . \tag{2.8}$$

In order to avoid cumbersome expressions, we now introduce two types of simplifying notation. First, integrations (usually but not necessarily over 4-space) will be represented by colons. Then

$$\langle 0_+(S)0_-\rangle = 1 + \tfrac{1}{2}iS:\Delta_+:S \tag{2.9}$$

Then we introduce a diagrammatic representation by denoting a source by ○ and the causal propagator by —◄— :

$$\langle 0_+(S)0_-\rangle = 1+\tfrac{1}{2}i(\bigcirc\!\!-\!\!\blacktriangleleft\!\!-\!\!\bigcirc) \ . \tag{2.10}$$

The next step is to construct the amplitudes for the generation or annihilation of any number of particles. Elements capable of operating in this fashion are called <u>strong sources</u>, and are regarded as being composed of a sufficiently large number of independent weak sources. Uniformity once again imposes a constraint on the form of the source amplitudes and gives an opportunity to introduce particle indistinguishability, and hence particle statistics, into the scheme.

In order to construct the strong source amplitudes, we express the vacuum persistence amplitude (v.p.a.) following the operation of a strong source S as the product of the v.p.a.'s. following operation of each of its weak components, $S = \sum_\alpha S_\alpha$:

$$\begin{aligned}\langle 0_+(S)0_-\rangle &= \prod_\alpha \{1 + \tfrac{1}{2}iS_\alpha:\Delta_+:S_\alpha\} \\ &\approx \exp\{\tfrac{1}{2}i\sum_\alpha S_\alpha:\Delta_+:S_\alpha\}\end{aligned} \tag{2.11}$$

But the v.p.a. cannot, by the principle of uniformity, depend on the individual components but only on the total source S. This is achieved if there is no coupling between the sources in the sense that all $S_\alpha:\Delta_+:S_\beta$ vanish identically for $\alpha\neq\beta$. Then

$$\langle 0_+(S)0_-\rangle = \exp\{\tfrac{1}{2}i\sum_{\alpha\beta} S_\alpha:\Delta_+:S_\beta\} = \exp\{\tfrac{1}{2}iS:\Delta_+:S\} \tag{2.12}$$

Diagrammatically we identify S as ●, the union of all the S_α, and so

$$\langle 0_+(S)0_-\rangle = \exp\{\tfrac{1}{2}i(\bullet\!\!-\!\!\blacktriangleleft\!\!-\!\!\bullet)\} \ . \tag{2.13}$$

Causality is now invoked again. Consider a sequence of source activities in which one strong source $S^{(-)}$ precedes another $S^{(+)}$. After this sequence the vacuum persists with an amplitude

$$\langle 0_+(S)0_-\rangle = \exp\{\tfrac{1}{2}i\sum_{\alpha\beta} S_{(\alpha)}:\Delta_+:S_{(\beta)}\}$$

$$= \exp\{\tfrac{1}{2}i[S^{(+)}:\Delta_+:S^{(+)} + S^{(-)}:\Delta_+:S^{(-)}$$

$$+ S^{(+)}:\Delta_+:S^{(-)} + S^{(-)}:\Delta_+:S^{(+)}]\}$$

$$= \langle 0_+(S^{(+)})0_-\rangle\{\exp[iS^{(+)}:\Delta_+:S^{(-)}]\}\langle 0_+(S^{(-)})0_-\rangle. \qquad (2.14)$$

But because of the ordering of sources $t^{(+)} > t^{(-)}$, and so

$$iS^{(+)}:\Delta_+:S^{(-)} = i\int S^{(+)*}(P)iS^{(-)}(P)\,d\omega_P$$

$$= \sum_P iS_P^{(+)*}iS_P^{(-)} \; . \qquad (2.15)$$

It follows that

$$\langle 0_+(S)0_-\rangle = \langle 0_+(S^{(+)})0_-\rangle \prod_P \exp(iS_P^{(+)*}iS_P^{(-)})\langle 0_+(S^{(-)})0_-\rangle$$

$$= \langle 0_+(S^{(+)})0_-\rangle \prod_P \sum_{nP} \left\{ \frac{(iS_P^{(+)*})^{n_P}}{(n_P!)^{\frac{1}{2}}} \cdot \frac{(iS_P^{(-)})^{n_P}}{(n_P!)^{\frac{1}{2}}} \right\} \langle 0_+(S^{(-)})0_-\rangle.$$

(2.16)

The alternative approach to the source analysis is to regard the strong source as generating a superposition of n-particle states, which are subsequently annihilated by another strong source. The causal analysis is

$$\langle 0_+(S)0_-\rangle = \sum_{\{n\}} \langle 0_+(S^{(+)})\{n\}\rangle\langle\{n\}(S^{(-)})0_-\rangle \; . \qquad (2.17)$$

Comparison of the two forms gives expressions for the n-particle source amplitudes:

$$\langle 0_+(S)\{n\}\rangle = \langle 0_+(S)0_-\rangle \prod_P \{(iS_P^*)^{n_P}/(n_P!)^{\frac{1}{2}}\} \qquad (2.18a)$$

$$\langle\{n\}(S)0_-\rangle = \langle 0_+(S)0_-\rangle \prod_P \{(iS_P)^{n_P}/(n_P!)^{\frac{1}{2}}\} \; . \qquad (2.18b)$$

On introducing the function

$$\phi_{\{n\}}(x) = \prod_P \{(-ix_P)^{n_P}/(n_P!)^{\frac{1}{2}}\} \qquad (2.19)$$

one obtains a generating function for the source amplitudes acting either as generators, $\langle x(S)0_-\rangle$, or as annihilators, $\langle 0_+(S)x\rangle$:

$$\begin{aligned}\langle x(S)0_-\rangle &\equiv \sum_{\{n\}} \phi_{\{n\}}(x)\langle\{n\}(S)0_-\rangle \\ &= \sum_{\{n\}} \langle 0_+(S)0_-\rangle \prod_P \{(x_P S_P)^{n_P}/n_P!\} \\ &= \langle 0_+(S)0_-\rangle \exp(x,S) \qquad (2.20)\end{aligned}$$

where $(x,S) = \sum_P x_P S_P$. Similarly

$$\langle 0_+(S)x\rangle = \langle 0_+(S)0_-\rangle \exp(x,S^*) \ . \qquad (2.21)$$

The way in which the particle statistics enter the discussion can now be demonstrated by investigating the amplitude with which an additional source can inject particles into a region of space-time already occupied by particles. The causal analysis requires the introduction of a weak probe source, $S^{(0)}$, in the region of space-time lying between the strong sources $S^{(+)}$ and $S^{(-)}$. $S^{(0)}$ injects particles (or annihilates them); but does so only weakly. Therefore, one causal analysis of the overall v.p.a. is

$$\begin{aligned}\langle 0_+(S)0_-\rangle &= \sum_{\{n_\pm\}} \langle 0_+(S^+)\{n_+\}\rangle\langle\{n_+\}(S^{(0)})\{n_-\}\rangle\langle\{n_-\}(S^{(-)})0_-\rangle \\ &= \sum_{\{n_+\}} \{\langle 0_+(S^+)\{n_+\}\rangle\langle\{n_+\}(S^{(0)})\{n_++1\}\rangle\langle\{n_++1\}(S^{(-)})0_-\rangle \\ &\quad + \langle 0_+(S^+)\{n_+\}\rangle\langle\{n_+\}(S^{(0)})\{n_+-1\}\rangle\langle\{n_+-1\}(S^{(-)})0_-\rangle\} ,\end{aligned} \qquad (2.22)$$

where $\{n_+\pm 1\}$ signifies that one particle has been added to or subtracted from one of the available states. The term corresponding to the diagonal element of $S^{(0)}$ has been omitted. An alternative causal analysis is based on eqn (2.12):

$$\begin{aligned}\langle 0_+(S)0_-\rangle &= \langle 0_+(S^{(+)}+S^{(-)})0_-\rangle \exp(iS^{(+)}:\Delta_+:S^{(0)}) \\ &\quad \times \exp(iS^{(0)}:\Delta_+:S^{(-)})\langle 0_+(S^{(0)})0_-\rangle\end{aligned}$$

$$= \sum_{\{n\}} \langle 0_+(S^{(+)})\{n\}\rangle\langle\{n\}(S^{(-)})0_-\rangle\langle 0_+(S^{(0)})0_-\rangle$$

$$\times \exp\{ \sum_P is_P^{(+)*} is_P^{(0)} + \sum_P is_P^{(0)*} is_P^{(-)}\}$$

$$\approx \sum_{\{n\}} \langle 0_+(S^{(+)})\{n\}\rangle\langle\{n\}(S^{(-)})0_-\rangle\langle 0_+(S^{(0)})0_-\rangle$$

$$\times \{\sum_P is_P^{(+)*} is_P^{(0)} + \sum_P is_P^{(0)*} is_P^{(-)}\} \qquad (2.23)$$

if $S^{(0)}$ is weak and we omit the diagonal contribution of $S^{(0)}$ (the leading 1 of the expansion of the exponential). We now equate powers of $S^{(+)}$ and $S^{(-)}$ in the two expressions using the expressions for the strong source amplitudes. This gives, with $S^{(0)}$ written as S,

$$\langle (n+1)_P(S)n_P\rangle = n_P^{\frac{1}{2}}\, is_P^* \qquad (2.24a)$$

$$\langle (n-1)_P(S)n_P\rangle = (n_P+1)^{\frac{1}{2}} is_P \qquad (2.24b)$$

showing the underlying Bose-Einstein statistics. Fermi-Dirac statistics can also be obtained either by antisymmetrizing the sources (6) or by antisymmetrizing the causal propagator (5).

We have now established enough properties concerning sources to be able to move on to the next development: the determination of the influence of a source on a remote region of space-time. That is, we need to be able to deal with the fields generated by sources. The outcome of this step in the analysis is a route to the determination of explicit expressions for the source amplitudes.

Consider the effect on $\mathfrak{S} = \frac{1}{2}S{:}\Delta{:}S$ (which governs the v.p.a., eqn (2.12)) when a weak source is added to the existing system: it changes by an amount

$$\delta\mathfrak{S} = \delta S{:}\Delta_+{:}S = \int dR dR' \delta S(R)\Delta_+(R-R')S(R')$$

$$= \int dR \phi(R)\delta S(R) = \phi : \delta S \qquad (2.25)$$

where $\phi(R) = \int dR' \Delta_+(R-R')S(R') = \Delta_+{:}S \,. \qquad (2.26)$

Since we already know the equation satisfied by Δ_+, eqn (2.8), it follows that ϕ satisfies

$$(-\partial^2 + m^2)\phi(R) = S(R), \tag{2.27}$$

which identifies it as the <u>field</u> arising from the source S. Through ϕ we are able to express the effect at some point R of the source active in some region of space-time. The diagrammatic representation of a field is —◀—● if it arises from a strong source, and —◀—○ if it arises from a weak source.

At this stage we have at our disposal various ways of expressing the quadratic form $\mathfrak{S}$ appearing in the v.p.a.:

$$\mathfrak{S} = \tfrac{1}{2}S{:}\Delta_+{:}S = \tfrac{1}{2}S{:}\phi = \tfrac{1}{2}\phi{:}(-\partial^2 + m^2)\phi \ . \tag{2.28}$$

We are therefore free to take any linear combination, and in particular to take

$$\mathfrak{S} = S{:}\phi - \tfrac{1}{2}\phi{:}(-\partial^2 + m^2)\phi, \tag{2.29}$$

If S and ϕ are temporarily regarded as independent variables, a variation of $\mathfrak{S}$ is

$$\delta\mathfrak{S} = \delta S{:}\phi + S{:}\ \delta\phi - \delta\phi{:}(-\partial^2 + m^2)\phi \tag{2.30}$$

In order for this to be compatible with eqn (2.25) it follows that $\mathfrak{S}$ must be stationary with respect to variations in the fields, the implication then being that $S{:}\delta\phi - \delta\phi{:}(-\partial^2 + m^2)\phi$ must vanish. This in turn implies eqn (2.27). Since a stationarity principle applied to furnishes the dynamical equation of motion for the system, it follows that $\mathfrak{S}$ has been established as being the <u>action</u> of the system. In other words, the v.p.a. can now be expressed in terms of the system's action through

$$\langle 0_+(S)0_-\rangle = \exp(\tfrac{1}{2}i\mathfrak{S}) \tag{2.31}$$

which in principle is known (e.g., it may be expressed in terms of a lagrangian).

The penultimate step in this account of the basic formalism is to set up the technique for allowing fields to interact, for this is what is required when a field of molecules (generated and annihilated by some causal arrangement of sources) coincides in some region of space-time with a field of photons (in spectroscopy) or of other molecules (in reactive or unreactive collision processes) generated by some other appropriately organized causal sequence of sources. We suppose that the interaction is bilinear in the two fields ϕ, ϕ' arising from the two types of source S, S', and therefore that it contributes a term

$\phi{:}I{:}\phi'$ to the interaction. This is not entirely general : it excludes, for instance, the A^2 contribution from the vector potential of an electromagnetic field. Nevertheless, the restriction keeps the development simple, and the generalization to a more general case is straightforward.

In the presence of an interaction the action is

$$\overline{\mathfrak{S}} = \mathfrak{S} + \mathfrak{S}' + \phi{:}I{:}\phi' , \qquad (2.32)$$

where $\mathfrak{S}$ and $\mathfrak{S}'$ are the actions for the individual fields, and therefore have the form given in eqn (2.28). It is convenient to write $-\partial^2 + m^2$ as the operator Ω, and $-\partial^2 + m'^2$ as Ω', for then

$$\overline{\mathfrak{S}} = S{:}\phi + S'{:}\phi' - \tfrac{1}{2}\phi{:}\ \Omega\phi - \tfrac{1}{2}\phi'{:}\Omega'\phi' + \phi{:}I{:}\phi' . \qquad (2.33)$$

The two coupled field equations of motion come from the variation of $\overline{\mathfrak{S}}$ with respect to ϕ and ϕ', and the stationarity condition. This gives

$$\delta_\phi\overline{\mathfrak{S}} = S - \Omega\phi + I{:}\phi' = 0 \Rightarrow \Omega\phi = S + I{:}\phi' \qquad (2.33)$$

$$\delta_{\phi'}\overline{\mathfrak{S}} = S' - \Omega'\phi' + \phi{:}I = 0 \Rightarrow \Omega'\phi' = S' + \phi{:}I \qquad (2.34)$$

These two field equations can be solved by finding the solution of the modified propagator equations

$$\{\Omega - (I{:}\phi')\}\overline{\Delta}_+ = \delta \quad \text{and} \quad \{\Omega' - (I{:}\phi)\}\overline{\Delta}'_+ = \delta \qquad (2.35)$$

(where δ is a delta-function), for then

$$\phi = \overline{\Delta}_+ {:}S \qquad \text{and} \qquad \phi' = \overline{\Delta}'_+ {:}S' \qquad (2.36)$$

are the interacting fields. The presence of the S' source affects the propagation of the field from the S source, and vice-versa, because each appears in the equation for the other's propagator. The general solutions can be found only by setting up some kind of perturbation expansion. Thus, if Δ_+ is the solution of the equation $\Omega\Delta_+ = \delta$ (and Δ'_+ the solution of $\Omega'\Delta'_+ = \delta$), then it follows that formal solutions are

$$\overline{\Delta}_+ = \Delta_+ + \Delta_+ {:}(I{:}\phi')\overline{\Delta}_+ \quad \text{and} \quad \overline{\Delta}'_+ = \Delta'_+ + \Delta'_+{:}(I{:}\phi)\overline{\Delta}'_+ \qquad (2.37)$$

(use $\delta{:}f = f$). Both expressions can be iterated in powers of the interaction or, equivalently, in terms of the order of the other source (ϕ' is first-order in S'):

$$\begin{aligned}\overline{\Delta}_+ &= \Delta_+ + \Delta_+{:}(I{:}\phi')\Delta_+ + \Delta_+{:}(I{:}\phi')\Delta_+{:}(I{:}\phi')\Delta_+ + \ldots\\ \overline{\Delta}'_+ &= \Delta'_+ + \Delta'_+ {:}(I{:}\phi)\Delta'_+ + \Delta'_+{:}(I{:}\phi)\Delta'_+ {:}(I{:}\phi)\Delta'_+ + \ldots.\end{aligned} \qquad (2.38)$$

We are interested in the action itself, because the matrix elements of the sources come from it _via_ the v.p.a.; therefore, the solutions in eqns (2.37) are substituted into eqn (2.33):

$$\overline{\mathfrak{S}} = S{:}\phi + S'{:}\phi' - \tfrac{1}{2}\phi{:}\{S+I{:}\phi'\} - \tfrac{1}{2}\phi'{:}\{S'+I{:}\phi\} + \phi{:}I{:}\phi'$$

$$= \tfrac{1}{2}S{:}\phi + \tfrac{1}{2}S'{:}\phi'$$

$$= \tfrac{1}{2}S{:}\overline{\Delta}_+{:}S + \tfrac{1}{2}S'{:}\overline{\Delta}'_+{:}S'. \qquad (2.39)$$

On substituting the dressed propagator expansions (eqn (2.38)) into eqn (2.39) we can collect terms corresponding to a particular number of source appearances:

$$\overline{\mathfrak{S}} = \sum_{N,N'} \mathfrak{S}(N,N') \ . \qquad (2.40)$$

In subsequent applications we shall often deal with the case in which N' = 2, N indefinite. For instance, in an electromagnetic interaction we prepare and later destroy a molecule, and in the intermediate space-time region allow it to interact with a number of photons. This means that the partial actions of interest are $\mathfrak{S}(N,2)$. These have the form

$$\mathfrak{S}(0,2) = \tfrac{1}{2}S'{:}\Delta'_+{:}S' \qquad (2.41a)$$

$$\mathfrak{S}(1,2) = \tfrac{1}{2}S'{:}\Delta'_+{:}(I{:}\phi)\Delta'_+{:}S' \qquad (2.41b)$$

$$\mathfrak{S}(2,2) = \tfrac{1}{2}S{:}\Delta_+{:}(I{:}\phi')\Delta_+{:}(I{:}\phi')\Delta_+{:}S + \tfrac{1}{2}S'{:}\Delta'_+{:}(I{:}\phi)\Delta'_+{:}(I{:}\phi)\Delta'_+{:}S' \qquad (2.41c)$$

and so on (some of the terms vanish because of the causality sequence; for example, the first term in $\mathfrak{S}(2,2)$ vanishes if we ensure that the molecule sources operate outside the space-time region occupied by the photons - that is, the photons interact with an existing molecule).

Before showing the use to which the $\mathfrak{S}(N,N')$ are put, we identify the general focus of interest in an experiment. The v.p.a. contains information about all the paths taken between the emission and absorption zones. The causal analysis can be expressed in terms of the initial generation of independent particles, their propagation into a region of space-time where they may interact, and then their annihilation by a subsequent source. This sequence can be expressed in terms of transition amplitudes $\langle\{m\}|\{n\}\rangle$ between the generated and annihilated set of states. Consequently,

$$\langle 0_+(\overline{S}^{(+)}+\overline{S}^{(-)})0_-\rangle = \sum_{\{m,n\}} \langle 0_+(\overline{S}^{(+)})\{m\}\rangle\langle\{m\}|\{n\}\rangle\langle\{n\}(\overline{S}^{(-)})0_-\rangle \qquad (2.42)$$

In this expression $\overline{S}$ represents a composite source (S + S') and the occupation numbers relate to total states, the direct products of the particles produced by both S and S', $|\{n\}\rangle \equiv |\{n\}\rangle \otimes |\{n'\}\rangle$. Their amplitudes are

$$\langle 0_+(\overline{S})\{n\}\rangle = \langle 0_+(S)\{n\}\rangle\langle 0_+(S')\{n'\}\rangle$$

$$= \langle 0_+(\overline{S})0_-\rangle \prod_{PP'} \left\{\frac{(iS_P^*)^{n_P}}{(n_P!)^{\frac{1}{2}}}\right\}\left\{\frac{(iS'^*_{P'})^{n'_{P'}}}{(n'_{P'}!)^{\frac{1}{2}}}\right\} \qquad (2.43a)$$

$$\langle\{n\}(\overline{S})0_-\rangle = \langle 0_+(\overline{S})0_-\rangle \prod_{PP'} \left\{\frac{(iS_P)^{n_P}}{(n_P!)^{\frac{1}{2}}}\right\}\left\{\frac{(iS'_{P'})^{n'_{P'}}}{(n'_{P'}!)^{\frac{1}{2}}}\right\}, \qquad (2.44b)$$

which implies that eqn (2.42) can be treated as a power series in the sources:

$$\langle 0_+(\overline{S}^{(+)} + \overline{S}^{(-)})0_-\rangle = \langle 0_+(\overline{S}^{(+)})0_-\rangle\langle 0_+(\overline{S}^{(-)})0_-\rangle$$

$$\times \sum_{\substack{n',n\\ m',m}} \prod_{PP'} \prod_{RR'} \langle\{m\}|\{n\}\rangle \left\{\frac{(iS_P^*)^{m_P}}{(m_P!)^{\frac{1}{2}}}\cdot\frac{(iS'^*_{P'})^{m_{P'}}}{(m_{P'}!)^{\frac{1}{2}}}\cdot\frac{(iS_R)^{n_R}}{(n_R!)^{\frac{1}{2}}}\cdot\frac{(iS'_{R'})^{n'_{R'}}}{(n'_{R'}!)^{\frac{1}{2}}}\right\}.$$

(2.45)

But we already have a power series in the sources which is explicit in the interaction between the two fields, since we know that the v.p.a. for the causal sequence under discussion is nothing other than $\exp i\mathfrak{S}$, and $\mathfrak{S}$ is itself given as a power series in the source amplitudes. Now we come to a subtle point: it is not necessary to deal with $\exp i\mathfrak{S}$ itself, for the powers of $\mathfrak{S}$ merely correspond to various repetitions of the basic interaction (4). $i\mathfrak{S}$ itself is in fact the selection of all <u>irreducible</u> interaction processes. Therefore all we have to do is to write

$$i\mathfrak{S} = \sum_{\{m,n\}} \langle 0_+(\overline{S}^{(+)})\{m\}\rangle\langle\{m\}|\{n\}\rangle\langle\{n\}\,\overline{S}^{(-)})0_-\rangle \qquad (2.46)$$

with the l.h.s. expressed as a sum of $\mathfrak{S}(N,N')$ for various powers of both sources and the r.h.s. as in eqn (2.45), and then identify the transition elements by comparing powers of the source amplitudes. In this way we arrive at the central quantities of interest: the transition matrix elements between the initial and final states. Note, too, that the source amplitudes cancel at this stage: they have merely been a vehicle for the development. Once we have the transition matrix elements we have, in effect, the $\mathbb{T}$-matrix elements of $\mathbb{S}$-matrix theory.

3. EXPLICIT ACTIONS

At this stage it is necessary to derive the source structure of two kinds of system: composite particles (such as molecules) with discrete eigenvalue spectra, and the electromagnetic field.

The partial Fourier transform of the causal propagator is

$$\Delta_+(R-R') = (1/2\pi)\int dP^o \Delta_+(\vec{r}-\vec{r}', P^o)\exp\{-iP^o(t-t')\} \tag{3.1}$$

and if the eigenfunctions of the composite particles are $\psi_{P^{o'}}(\vec{r})$, then

$$\Delta_+(\vec{r}-\vec{r}', P^o) = \sum_{P^{o'}} \left\{ \left(\frac{\psi_{P^{o'}}(\vec{r})\psi^*_{P^{o'}}(\vec{r}')}{P^{o'} - P^o - i\eta} \right) + \left(\frac{\psi^*_{P^{o'}}(\vec{r})\psi_{P^{o'}}(\vec{r}')}{P^{o'} + P^o - i\eta} \right) \right\}. \tag{3.2}$$

Other labels, and the corresponding summation, have to be added if the functions are degenerate. It is readily checked, by contour integration, that $\Delta_+(R-R')$ vanishes for $t < t'$ and positive energies ($P^{o'}$), and for $t' < t$ and negative energies, and therefore has the correct causal structure. Explicit integration leads to

$$\Delta_+(R-R') = i\theta(t'-t)\sum_{P^{o'}} \psi^*_{P^{o'}}(\vec{r})\psi_{P^{o'}}(\vec{r}')e^{-iP^{o'}(t'-t)}$$

$$+\, i\theta(t-t')\sum_{P^{o'}} \psi_{P^{o'}}(\vec{r})\psi^*_{P^{o'}}(\vec{r}')e^{-iP^{o'}(t-t')} \tag{3.3}$$

The <u>action</u> is therefore

$$\begin{aligned} \mathfrak{S} &= \tfrac{1}{2}\, S:\Delta_+:S \\ &= \tfrac{1}{2}\int dR\,dR'\,S(R)\Delta_+(R-R')S(R') \\ &= \tfrac{1}{2}i\sum_{P^o}\int dt\,dt'\,dr\,dr'\,S(R)S(R')\{\psi^*_{P^o}(\vec{r})\psi_{P^o}(\vec{r}')\theta(t'-t)e^{-iP^o(t'-t)} \\ &\quad + \psi_{P^o}(\vec{r})\psi^*_{P^o}(\vec{r}')\theta(t-t')e^{-iP^o(t-t')}\} \end{aligned}$$

$$= \tfrac{1}{2}i \sum_{po} \int dt dt' \{ S_{po}(t) S^*_{po}(t') \theta(t'-t) e^{-iP^o(t'-t)}$$

$$+ \; S^*_{po}(t) S_{po}(t') \theta(t-t') e^{-iP^o(t-t')} \} \tag{3.4}$$

where $$S_{po}(t) \; = \; \int d\vec{r} S(R) \psi^*_{po}(\vec{r}) \; . \tag{3.5}$$

The field associated with the source is

$$\Psi \; = \; \Delta_+ : S \; = \; \int dR' \Delta_+(R-R') S(R')$$

$$= \; i \sum_{po} \int dr' dt' \{ \theta(t'-t) \psi^*_{po}(\vec{r}) \psi_{po}(\vec{r}') S(R') e^{-iP^o(t'-t)}$$

$$+ \; \theta(t-t') \psi_{po}(\vec{r}) \psi^*_{po}(\vec{r}') e^{-iP^o(t-t')} \}$$

$$= \; i \sum_{po} \int dt' \{ \theta(t'-t) \psi^*_{po}(\vec{r}) S^*_{po}(t') e^{-iP^o(t'-t)}$$

$$+ \; \theta(t-t') \psi_{po}(\vec{r}) S_{po}(t') e^{-iP^o(t-t')} \}$$

$$= \; i \sum_{po} \{ \psi^*_{po}(\vec{r}) S^*_{po} \, e^{iP^o t} + \psi_{po}(\vec{r}) S_{po} e^{-iP^o t} \} \tag{3.6}$$

where

$$S_{po} \; = \; \int dt \; S_{po'}(t) e^{iP^o t} \tag{3.7}$$

and we have taken note of the causal arrangement, such that the emission source (S) precedes absorption (S^*), and we are interested in the field at time intermediate between the two events ($t' > t$ for S^*, $t' < t$ for S). The field structure can therefore be written

$$\Psi(R) \; = \; \Psi^{(-)}(R) + \Psi^{(+)}(R) \tag{3.8}$$

where $\Psi^{(-)}(R)$ is the field after the source has operated, and $\Psi^{(+)}$ the field before it has operated (as a sink):

$$\psi^{(-)}(R) = \sum_{P^o} \psi_{P^o}(\vec{r})e^{-iP^o t}\, iS_{P^o} \tag{3.9a}$$

$$\psi^{(+)}(R) = \sum_{P^o} \psi^*_{P^o}(\vec{r})e^{iP^o t} iS^*_{P^o}. \tag{3.9b}$$

The other source structure we require is that of the electromagnetic field. Since this is a vector field we can anticipate that it is necessary to have a vector source: this we denote J. But it is well-known that the massless nature of the field requires special treatment, and in particular the spin-1 nature of the photon emerges from the 4-component (4-vector) source J if the constraint $\partial_\mu J^\mu = 0$ is imposed. This implies, incidentally, the existence of a conservation law (the conservation of electric charge). Since we are now dealing with massless particles, we have to use a special form of the propagator; therefore we write the v.p.a. under the impact of a sequence of photon sources as

$$\langle 0_+(J)0_-\rangle = \exp\{\tfrac{1}{2}iJ^\mu{:}D_+{:}J_\mu\} \tag{3.10}$$

where D_+ is the causal massless propagator, having the form

$$D_+(R-R') = (i/4\pi^2)\left(\frac{1}{|\vec{r}-\vec{r}'|}\right)\int_0^\infty dP^o\, \sin\{P^o|r-r'|\}\exp\{-P^o i|t-t'|\}. \tag{3.11}$$

The <u>action</u> for the electromagnetic field is

$$\mathcal{J} = \tfrac{1}{2}J^\mu{:}D_+{:}J_\mu \qquad \text{subject to} \quad \partial_\mu J^\mu = 0, \tag{3.12}$$

and the <u>field</u> will be denoted A (and identified with the vector potential of conventional electrodynamics):

$$A^\mu = D_+{:}J^\mu\ . \tag{3.13}$$

This field can be decomposed into causal components $A^{(\pm)}$ just as in the case of the field of the composite particle, and we write

$$\begin{aligned} A(R) &= A^{(-)}(R) + A^{(+)}(R) \\ &= \sum_{K,\lambda}\{e_{K\lambda}\, e^{iK\cdot R}\, iJ^{(-)}_{K\lambda} + e_{K\lambda}e^{-iK\cdot R} iJ^{(+)*}_{K\lambda}\} \end{aligned} \tag{3.14}$$

where K is the momentum 4-vector and λ denotes a polarization state of the field ($K\cdot R = \vec{k}\cdot\vec{r} - K^{o}t$). The explicit form for the action is the one that leads to the conventional equations of motion for the electromagnetic field:

$$\mathcal{J} = J^{\mu}:A_{\mu} + L[A] \tag{3.15}$$

where the lagrangian functional of A is

$$L[A] = -\tfrac{1}{4}F^{\mu\nu}:F_{\mu\nu} \quad , \qquad F_{\mu\nu} = \partial_{\mu}A_{\nu} - \partial_{\nu}A_{\mu} . \tag{3.16}$$

Finally, we need the explicit form for the action when both the composite system sources and the electromagnetic field sources generate fields in a common region of space-time. From the discussion in §2 we can expect to express the joint action in terms of the dressed propagator:

$$\mathcal{W} = J^{\mu}:A_{\mu} - \tfrac{1}{4}F_{\mu\nu}:F^{\mu\nu} + S:\overline{\Delta}_{+}:S \quad , \tag{3.17}$$

where $\overline{\Delta}_{+}$ is the solution of the coupled equation

$$\{(H - i\partial_{t}) + ie(\partial\cdot A+A\cdot\partial) + e^{2}A\cdot A\} \ \overline{\Delta}_{+} = \delta . \tag{3.18}$$

That this is so is shown in the Appendix. It is also possible to express $\mathcal{W}$ in a form which is based on $J:D_{+}:J$, but our interest will centre on the case where a single molecule (that is, a 2-source arrangement) interacts with an arbitrary number of photons, and the form of the action quoted above is most appropriate. The other form is appropriate when one wishes to emphasize that particle-particle interactions are mediated by photons.

In the notation of §2 the interaction term $I:\phi$ is now to be identified as

$$I:\phi = (e/m)\vec{p}\cdot\vec{A} - (e^{2}/2m)A^{2} , \tag{3.19}$$

and so the dressed propagator expands as follows:

$$\overline{\Delta}_{+} = \Delta_{+} + \Delta_{+}:\{+(e/m)\vec{p}\cdot\vec{A} - (e^{2}/2m)A^{2}\}:\Delta_{+} + \ldots \tag{3.20}$$

It follows that the particle field is

$$\begin{aligned} \Psi &= \overline{\Delta}_{+}:S \\ &= \Delta_{+}:S + \Delta_{+}:\{(e/m)\vec{p}\cdot\vec{A} - (e^{2}/2m)A^{2}\}:\Delta_{+}:S + \ldots \end{aligned} \tag{3.21}$$

4. BASIC APPLICATIONS

The analysis of any absorption or scattering process takes note of three steps. Initially a molecule and a photon are generated independently by $S^{(-)}$ and $J^{(-)}$. Then the two component fields propagate from the sources into the interaction zone, where the absorption or scattering process takes place. Finally, the component fields are eliminated by a pair of independent sources $S^{(+)}$ and $S^{(-)}$ acting as sinks. The expansion of $\bar{\Delta}_+$, or the dressed field, is equivalent to an expansion in powers of the number of photon sources in operation. The basic process of this kind contributes the term $\mathcal{W}(2,2)$ to the expansion of the total action, and so

$$\begin{aligned}\mathcal{W}(2,2) &= \tfrac{1}{2}(S:\bar{\Delta}_+:S)_{(2,2)} \\ &= \tfrac{1}{2}\Psi(e/m)\vec{p}.\vec{A}:\Delta_+:(e/m)\vec{p}.\vec{A}\Psi - \tfrac{1}{2}(e^2/2m)\Psi:A^2\Psi \end{aligned} \qquad (4.1)$$

Comparing this expression with

$$i\mathcal{W} = \sum_{i,f} <0_+(S^{(+)}J^{(+)})f><f|i><i(S^{(-)}J^{(-)})0_-> \qquad (4.2)$$

based on the same causal sequence lets us identify the transition amplitudes $<f|i>$.

The analysis of the second form leads to

$$i\mathcal{W}(2,2) = \sum_{i,f} iS_f^{(+)*}iJ_f^{(+)*}iS_i^{(-)}iJ_i^{(-)}<f|i>$$

or

$$\mathcal{W}(2,2) = -i\sum_{i,f} S_f^{(+)*}S_i^{(-)}J_f^{(f)*}J_i^{(-)}<f|i> \qquad (4.3)$$

where the i,f labels may refer to a variety of different characteristics (e.g. the pertinent state of the molecules when attached to S, and the wave-vector and polarization state of the photons when attached to J).

The analysis of the first expression for the partial action is based on the causal decomposition in eqns (3.8) and (3.14):

$$\Psi(R) = \psi_i(r)e^{-i\varepsilon_i t}iS_i^{(-)} + \psi_f^*(r)e^{i\varepsilon_f t}iS_f^{(+)*} \qquad (4.4)$$

$$A(R) = \vec{e}_i e^{iK_i\cdot R} iJ_i^{(-)} + \vec{e}_f e^{-iK_f\cdot R} iJ_f^{(+)*} \qquad (4.5)$$

where P^o has been written ε, the molecular eigenstate of interest. The relevant contribution to the partial action is therefore

$$\begin{aligned}\mathcal{W}(2,2) = & \sum_{i,f} iS_f^{(+)*} iS_i^{(-)} iJ_f^{(+)*} iJ_i^{(-)} \\ & \times \{(e/m)^2 \psi_f^* e^{i\varepsilon_f t} \vec{p}\cdot\vec{e}_f e^{-iK_f\cdot R} :\Delta_+: \vec{p}.\vec{e}_i e^{iK_i R'} \psi_i e^{-i\varepsilon_i t'} \\ & + (e/m)^2 \psi_f^* e^{i\varepsilon_f t} \vec{p}.\vec{e}_i e^{iK_i\cdot R} :\Delta_+: \vec{p}\cdot\vec{e}_f e^{-iK_f . R'} \psi_i e^{-i\varepsilon_i t'} \\ & - (e^2/m) \psi_f^* e^{i\varepsilon_f t} : (\vec{e}_f\cdot\vec{e}_i) e^{i(K_i - K_f)\cdot R} \psi_i e^{-i\varepsilon_i t}\}. \qquad (4.6)\end{aligned}$$

The transition amplitude $<f|i>$ can now be identified. Furthermore, we have the explicit form for the causal propagator in eqn (3.3), noting that $t > t'$ in the analysis. Since

$$e^{i\alpha t} : \theta(t-t') : e^{i\beta t'} = 2\pi i\delta(\alpha+\beta)/(\alpha + i\eta) = -2\pi i\delta(\alpha+\beta)/(\beta+i\eta) \qquad (4.7)$$

we can arrive at once (or almost at once) at

$$<f|i> = 2\pi i\delta(\varepsilon_f - \varepsilon_i + k_f - k_i)\{R_{fi}^{(1)} + R_{fi}^{(2)} + R_{fi}^{(3)}\} \qquad (4.8)$$

where

$$R_{fi}^{(1)} = (e/m)^2 \sum_g \left\{\frac{(f,\vec{p}.\vec{e}_f e^{-i\vec{k}_f.\vec{r}} g)(g,\vec{p}.\vec{e}_i e^{i\vec{k}_i.\vec{r}} i)}{\varepsilon_g - \varepsilon_i - k_i - i\eta}\right\} \qquad (4.9a)$$

$$R_{fi}^{(2)} = (e/m)^2 \sum_g \left\{\frac{(f,\vec{p}.\vec{e}_i e^{i\vec{k}_i.\vec{r}} g)(g,\vec{p}.\vec{e}_f e^{-i\vec{k}_f.\vec{r}} i)}{\varepsilon_g - \varepsilon_i + k_f - i\eta}\right\} \qquad (4.9b)$$

$$R_{fi}^{(3)} = -(e^2/m)(f,(\vec{e}_f.\vec{e}_i)e^{i(\vec{k}_i-\vec{k}_f).\vec{r}}i) \qquad (4.9c)$$

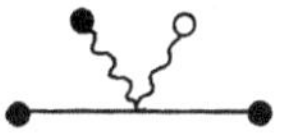

and the molecular eigenstates are ε_i, ε_f, ε_g. Note that an energy-conserving δ-function emerges automatically, and that there is a natural diagrammatic representation. Indeed, these expressions are the starting point of theories of optical birefringence, etc. (1-4).

5. CONCLUSION

We have shown that the two principles of causality and uniformity can be used to set up expressions for the scattering matrix in terms of the fundamental interaction hamiltonian components. From the $\mathbb{R}$-matrix elements in §4 it is a simple matter to deduce a wide range of optical properties, such as optical activity, natural and induced optical birefringence, and absorptive processes. The techniques can be quite readily extended to more complicated phenomena, and it is here that some of the practical advantages (as opposed to the purely intellectual advantages of seeing how two such fundamental principles can be used to circumscribe and discuss molecular properties) begin to emerge.

In the first place, the extension to n-photon interactions is quite straightforward. The simplest procedure is to note that the molecule interacting with the electromagnetic field is itself a source: it creates and annihilates photons, and adds and subtracts photons from beams of various constitutions. But we saw how the strength of a boson source varied with the occupation numbers of the states it was modifying: therefore we can use eqn (2.24) directly to write down the amplitudes, and hence the R-matrix elements. Essentially all this does is to introduce factors of $\sqrt{n}$ or $\sqrt{(n+1)}$, etc. (4).

The source technique can also very readily be applied to multiple-photon (non-linear) processes, for then all that is necessary is to extend the calculation to include higher-order partial actions. Thus two photon nonlinearities (like intensity-dependent optical activity (1), and the optical Faraday effect (10)) can be treated on the basis of $\mathcal{W}(4,2)$, which has the formal structure

$$\begin{aligned}\mathcal{W}(4,2) = {} & \Psi p.A:\Delta_+:p.A:\Delta_+:p.A:\Delta_+:p.A\Psi \\ & + \Psi p.A:\Delta_+:p.A:\Delta_+:A^2\Psi \\ & + \Psi p.A:\Delta_+:A^2:\Delta_+:p.A\Psi \\ & + \Psi A^2:\Delta_+:A^2\Psi . \end{aligned} \qquad (5.1)$$

(The diagrammatic analysis leads to 21 different terms.)

In the case of particle-particle interactions it is appropriate to start with the action in the alternative version mentioned in §3. Then one concentrates explicitly on the photon as the agent of the interaction, and the primitive interaction between two molecules in any state can be discussed in terms of $\mathcal{W}(0,4)$, for there the molecules act as the source of the photons carrying the interaction, and so it is unnecessary to introduce additional sources. Furthermore, by concentrating on the photon propagator from the outset, it turns out that a significant reduction in the number of diagrams takes place when compared with the same calculation done on the basis of S-matrix theory. Of course, the final result is the same, but source theory groups the diagrams together in a more systematic fashion.

Source theory has also been applied to fermion scattering (e.g. e^--H scattering). In this application the essential modification is to antisymmetrize the sources (or the fields) (5). The outcome of the calculation (as generally, but not always, elsewhere in source theory) are the transition matrix elements obtained in more conventional treatments.

ACKNOWLEDGEMENTS

The material reported here is based on work carried out mainly by Dr E.J. Austin (11) and also by Dr F.W. King, while both were members of this laboratory. I should like to thank Dr Austin for additional recent helpful discussions.

APPENDIX

The general form of the action in the presence of sources is

$$\mathfrak{S} = S{:}\psi^* + S^*{:}\psi - (\hbar^2/2m)\,\partial^\mu\psi^*{:}\partial_\mu\psi - \psi^*{:}V{:}\psi \tag{A1.1}$$

When an electromagnetic field is present $\mathfrak{S}$ changes to $\bar{\mathfrak{S}}$, which is obtained by making the replacement $p \to p + eA$:

$$\bar{\mathfrak{S}} = S{:}\psi^* + S^*{:}\psi - (\hbar^2/2m)\,\partial^\mu\psi^*{:}\partial_\mu\psi - \psi^*{:}V{:}\psi$$

$$+ (\hbar e/2mi)(A^\mu\psi^*{:}\partial_\mu\psi + A_\mu\psi{:}\partial^\mu\psi^*) - (e^2/2m)A^\mu\psi^*{:}A_\mu\psi. \tag{A1.2}$$

Variation with respect to ψ^* leads to an equation of motion:

$$\delta\bar{\mathfrak{S}} = S{:}\delta\psi^* - (\hbar^2/2m)\,\partial^\mu\delta\psi^*{:}\partial_\mu\psi - \delta\psi^*{:}V{:}\psi$$

$$+ (\hbar e/2mi)(A^\mu\delta\psi^*{:}\partial_\mu\psi + A_\mu\psi{:}\partial^\mu\delta\psi^*) - (e^2/2m)A^\mu\delta\psi^*{:}A_\mu\psi$$

$$= 0. \tag{A1.3}$$

Since $f{:}\partial^\mu\delta\psi^* = -\partial^\mu f{:}\delta\psi^*$, this corresponds to

$$-(\hbar^2/2m)\,\partial^2\psi + V{:}\psi - (e/2m)(A.p+p.A)\psi + (e^2/2m)A^2\psi = S. \tag{A1.4}$$

The dressed propagator is then defined as the solution of

$$\Omega\bar{\Delta}_+ = \delta, \tag{A1.5}$$

where Ω is the operator on the left of the preceeding equation. The alternative form for the action is obtained by adding the action for the electromagnetic field, and considering its variation with A.

REFERENCES

1. Atkins, P.W. and Barron, L.D.: 1968, Proc. Roy. Soc. A304, p. 303.

2. Atkins, P.W., and Woolley, R.G.: 1970, Proc. Roy. Soc. A314, p.251.

3. Atkins, P.W., and Wilson, A.D.: 1972, Molec. Phys. 24, p.33.

4. Atkins, P.W., and Austin, E.J.: 1976, Molec. Phys. 31, p.1621.

5. Atkins, P.W., and King, F.W.: 1978, Molec. Phys. 35, p.883.

6. Schwinger, J.: 1970, Particles, sources, and fields, Vol.1, 1973, Vol.2 (Addison-Wesley).

7. Schwinger, J.: 1969, Particles and sources, (Gordon and Breach).

8. Schwinger, J.: 1966, Phys. Rev., 152, p.1219.

9. Schwinger, J.: 1967, Phys. Rev. 158, p.1391.

10. Atkins, P.W., and Miller, M.H.: 1968, Molec. Phys. 15, p. 503.

11. Austin, E.J.: 1976, D.Phil. Thesis, Oxford.

SYMPOSIUM V. LARGE MOLECULES OF BIOLOGICAL IMPORTANCE

Chairman: P.-O. Löwdin

Quantum Chemistry Group, University of Uppsala,
Uppsala, Sweden

and

Quantum Theory Project, University of Florida,
Gainesville, Florida, U.S.A.

THE SIZE-CONSISTENCY PROBLEM FOR LARGE MOLECULES

Per-Olov Löwdin
Quantum Chemistry Group, Uppsala University
and
Quantum Theory Project, University of Florida

As chairman for the symposium on large molecules, I have been asked by the organizers to say a few words about a problem I think is important at the current stage of development. Using experience from solid state theory, I feel that the size-consistency problem is one of the most important ones, i.e. the fact that the cohesive energy of a crystal should be proportional to its volume V, whereas the cohesive energy of a periodic polymer ought to be proportional to its length L.

Let us start with a study of the electrostatic energy. For the sake of simplicity, we will consider a system with fixed nuclei g having the atomic numbers Z_g and the Hamiltonian

$$H = e^2 \sum_{g<h} \frac{Z_g Z_h}{R_{gh}} + \sum_i \left(\frac{p_i^2}{2m} - e^2 \sum_g \frac{Z_g}{r_{ig}}\right) + \sum_{i<j} \frac{e^2}{r_{ij}}, \tag{1}$$

where i = 1,2,...N is summed over the N electrons. The system is described by a density matrix Γ:

$$\Gamma = \Gamma(x_1, x_2, \ldots x_N | x_1', x_2', \ldots x_N') \tag{2}$$

having the trace 1, i.e. $\mathrm{Tr}\Gamma = 1$. Introducing the reduced density matrices[1] $\Gamma(x_1 x_2 | x_1' x_2')$ and $\gamma(x_1 | x_1')$, one obtains for the total energy

$$\langle H \rangle_{AV} = \mathrm{Tr}H\Gamma = e^2 \sum_{g<h} \frac{Z_{gh}}{R_{gh}} + \frac{1}{2m}\int p_1^2 \gamma(x_1|x_1')dx_1 - e^2 \sum_g Z_g \int \frac{\gamma(x_1|x_1)}{r_{1g}} dx_1$$
$$+ e^2 \iint \frac{\Gamma(x_1 x_2 | x_1 x_2)}{r_{12}} dx_1 dx_2, \tag{3}$$

where the individual terms - except for the kinetic energy - are <u>not</u> size-consistent. However, considering the fact that $\gamma(x_1) = \gamma(x_1|x_1)$ is essentially the electron density, one may now study the corresponding "classical" electrostatic energy:

K. Fukui and B. Pullman (eds.), Horizons of Quantum Chemistry, 193–195.

$$E_1 = e^2 \sum_{g<h} \frac{Z_g Z_h}{R_{gh}} - e^2 \sum_g Z_g \int \frac{\gamma(x_1)}{r_{1g}} dx_1 + \frac{e^2}{2} \iint \frac{\gamma(x_1)\gamma(x_2)}{r_{12}} dx_1 dx_2. \qquad (4)$$

By proper combination of terms[2)], it is easily shown that E_1 may be written as a "Madelung energy" corresponding to a system of point charges and a "correction energy" due to the extension of the ions. Both of these terms are size-consistent. It is then evident that the remaining electrostatic term

$$E_2 = e^2 \iint \frac{\Gamma(x_1x_2|x_1x_2) - \frac{1}{2}\gamma(x_1)\gamma(x_2)}{r_{12}} dx_1 dx_2 \qquad (5)$$

which contains the exchange and correlation effects in the system is also size-consistent. The danger is that, if the terms in the energy are not properly combined, one can get practically any results one wants, which is certainly meaningless.

Another aspect of the size-consistency problem is closely connected with the perturbation theory treatment of the system. Let us assume that the particles in the configuration space may be divided into two groups (a) and (b) having the Hamiltonians H_a and H_b, respectively, and the interaction term $V = \lambda_{ab} H_{ab}$, where λ_{ab} is an artificially introduced coupling constant. In such a case, one knows[3)] that, in the case when $\lambda_{ab} = 0$, the exact eigenfunction should reduce to the form $\Psi = \Psi_a \Psi_b$, whereas the energy should take the form

$$E = E_a + E_b \qquad (6)$$

As is shown by the linked-cluster theorem in diagrammatic perturbation theory, it is by no means trivial to write the perturbation expansions in such a way that this result becomes obvious. If this is not done, the perturbation expansions will contain spurious terms which are not size--consistent, i.e. which are proportional to V^2, V^3,... or L^2, L^3,..., respectively. In molecular theory, these problems have recently been studied by Bartlett.

There is one more similarity between solids and polymers which I would like to emphasize. If the solid is not an insulator, the ground state is situated at the bottom of a conduction band, which means that there are many low-lying electronic states close to the ground state, which is hence approximately electronically degenerate. In a polymer with many single bonds, there is a very large number of conformations close to the true ground state, and the large molecule shows hence a "nuclear degeneracy" of the ground state. In both cases, the ground state is approximately degenerate, and it may then turn out to be more practical to describe such states by means of density matrices Γ instead of wave functions.

In conclusion, I would like to draw attention to a curve which I draw at the Vålådalen symposium in 1958 about the relation between the refinement of a theory and the agreement between theory and experiment. Usually a rough theory based on a particular model may be developed to give an almost 100% agreement with the experimental experience - one reaches what was then called the "Pauling point". However, if the theory is further refined - e.g. by introduction of the ion-core electrons, by considering the overlap integrals, or by inclusion of triply excited states, etc. - the agreement usually goes down, and one reaches a minimum which was then called the "Ph.D. point". Further refinement of the theory beyond this point may then hopefully lead to a better agreement, and so on, but the 100% agreement is seldom restored again.

Anyway, good agreement between theory and experiment is a necessary but not sufficient criterion that a theory is good. Many simple interpolation schemes give sometimes excellent results - we may recall that Mendelejeff interpolated properties of unknown elements in the periodic system - and interpolation schemes having a few parameters are highly useful in science without any reference to more deep-lying theories. In between, one has the semi-empirical theories which use certain experimental data to predict other experimental results. When it comes to large molecules, one has for a long time resorted to semi-empirical methods. However, in the symposium we start today, most of the speakers will report about results for large molecules obtained by quantum-mechanical ab-initio methods. This is a remarkable development due to the increase of the power and quality of the electronic computers and their software as well as of the general program technique. Without further ado, I will now call on the first speaker.

REFERENCES

1. Husimi, K.: 1940, Proc. Phys.-Math. Soc. Japan 22, pp. 240;
Löwdin, P.O.: 1955, Phys. Rev. 97, pp. 1474;
McWeeny, R.: 1955, Proc. Roy. Soc. (London) A232, pp. 114.

2. Löwdin, P.O.: 1956, Adv. Phys. 5, pp. 1, particularly p. 12.

3. Löwdin, P.O.: 1962, J. Math. Phys. 3, pp. 1171.

THEORETICAL STUDIES OF THE STRUCTURE OF HEME MODELS

A. Veillard, A. Dedieu and M.-M. Rohmer
E.R. 139 du C.N.R.S., Institut Le Bel,
Université L. Pasteur, Strasbourg, France

INTRODUCTION

Hemoglobin is the vital protein that conveys oxygen from the lungs to the tissues [1]. A hemoglobin molecule is made up of four subunits, with each subunit comprising a polypeptide chain, called the globin, and a heme (Fig. 1). The heme is an iron(II) protoporphyrin IX (Fig. 2). In the deoxy state of hemoglobin (namely deoxyhemoglobin or deoxyHb) the five-coordinated iron atom is high-spin (S = 2) and protrudes above the porphyrin (Fig. 3). This iron is linked to the protein through its fifth ligand, an imidazole molecule of a histidine residue, called the proximal histidine. The sixth coordination site of the iron atom in deoxyhemoglobin has a histidine (called the distal histidine) too far away to coordinate with the iron. DeoxyHb can combine reversibly with molecular oxygen to give oxyhemoglobin (oxyHb) :

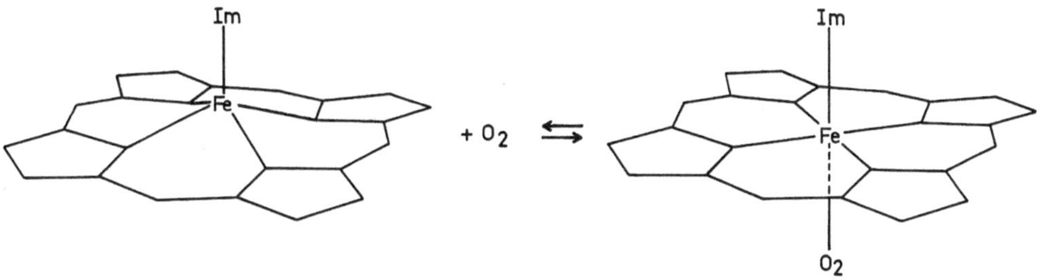

In oxyhemoglobin, molecular oxygen is attached to the iron atom as the sixth ligand. The six-coordinated iron is low-spin (S = 0) and is believed to be in the porphyrin plane.

K. Fukui and B. Pullman (eds.), Horizons of Quantum Chemistry, 197–225.

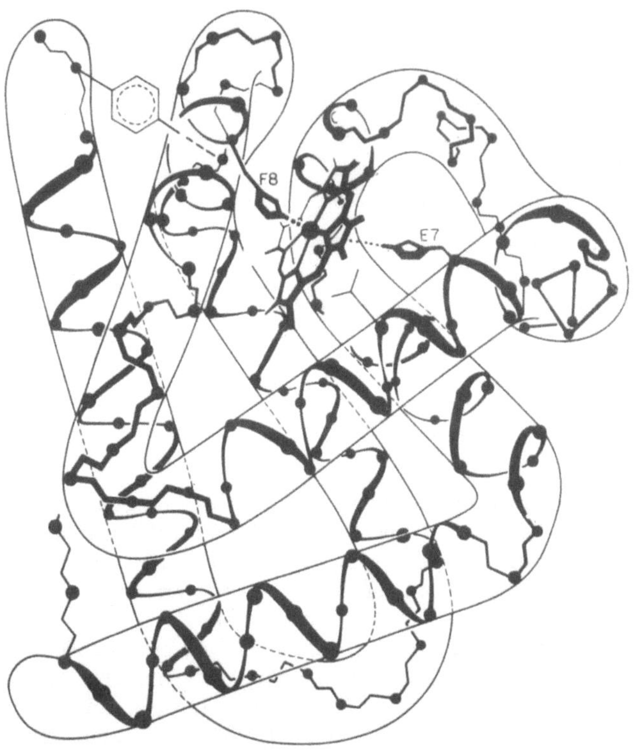

Fig. 1 A schematic representation of one of the four subunits of hemoglobin. The continuous blanck band represents the peptide chain. The heme group can be seen at the upper right center with the iron as a large dot. The coordinated histidine residue is labelled F8, the distal histidine residue E7 (reproduced from M. Perutz).

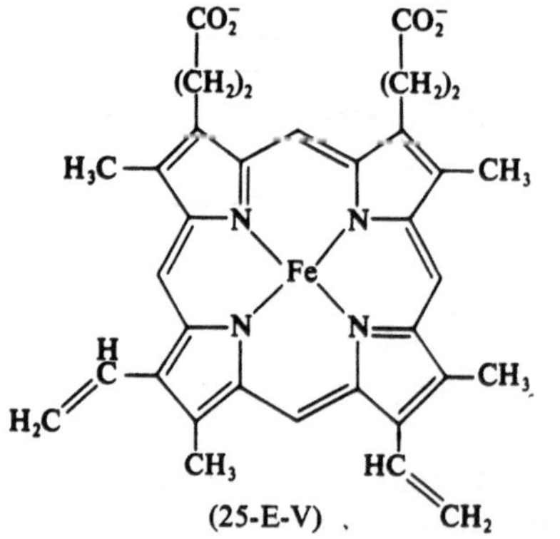

(25-E-V)

Fig. 2 The iron(II) protoporphyrin IX.

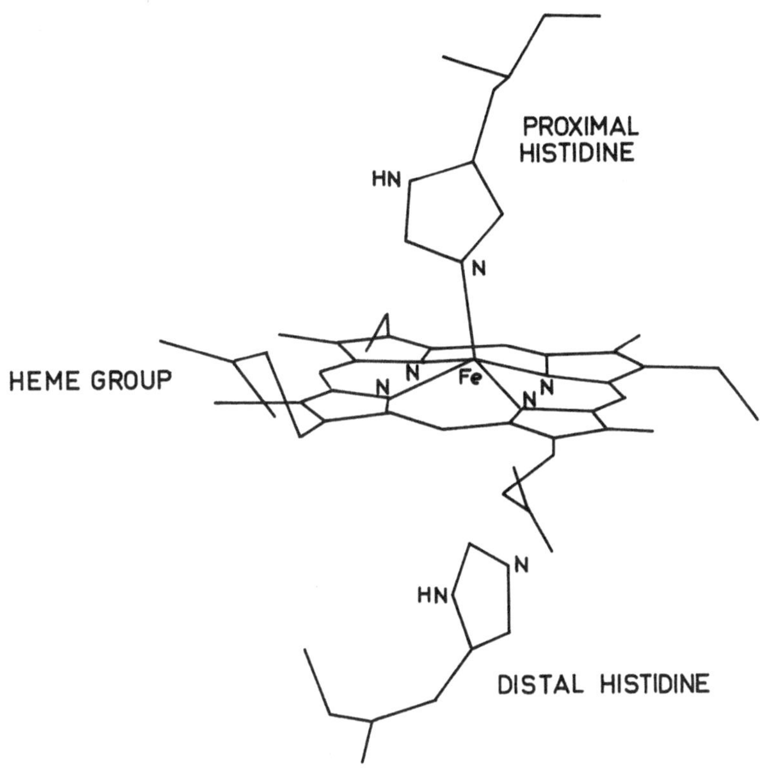

Fig. 3 Chemical structure of the heme group with the surrounding histidine residue in deoxyHb (reproduced from M. Perutz).

Hemoglobin displays rather unusual properties as an oxygen carrier in the sense that its affinity for oxygen rises with increasing oxygen saturation (the equilibrium curve of the percentage of oxygen bound as a function of the partial pressure of oxygen is sigmoid in shape rather than hyperbolic). This has been called the cooperative effects and represents a property of the hemoglobin tetramer which hinges on heme-heme interactions. These cooperative effects arise from an equilibrium between two structures, one characteristic for deoxyHb (T for tense) and the other for oxyHb (R for relaxed). According to a proposal by Perutz known as the trigger mechanism, the low-affinity T state is characterized by a large iron-porphyrin distance (of the order of 0.6 Å in deoxyHb) induced by the protein pulling on the proximal histidine, whereas the high-affinity R state is characterized by a smaller iron-porphyrin distance, resulting from a relaxation of the protein and a movement of the proximal histidine toward the heme. In the R state the iron can readily bind an oxygen molecule.

The structure and electronic properties of the heme are not yet completely understood. For instance the structure of the iron-dioxygen unit in oxyHb is not precisely known since the X-ray crystallographic studies of oxyHb have not located the oxygen atoms. Until recently it had not been possible to choose between the two structures proposed respectively by Pauling [2] (end-on structure 1) and Griffith [3] (side-on structure 2). In 1936, Pauling and Coryell found oxyHb to be diamagnetic

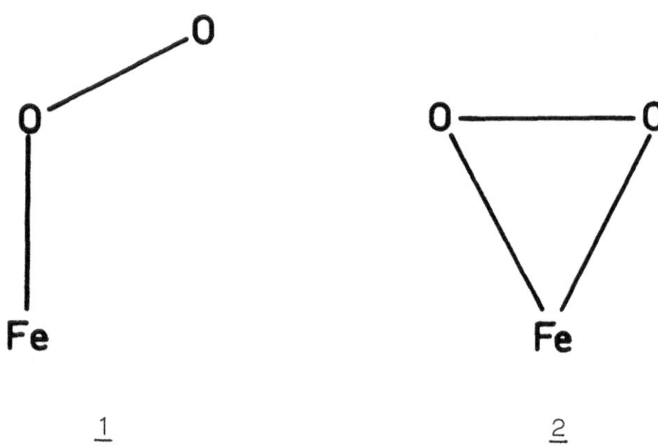

at 20°C [4], an observation which implies an electronic structure with all the electrons spin-paired in the heme group. However, fourty years later, measurements of the magnetic susceptibility of frozen aqueous solutions of human oxyhemoglobin in the temperature range 25-250°K indicated a temperature dependent behavior typical of a thermal equilibrium between a singlet ground state and a triplet excited state having two unpaired electrons per heme group, with an energy separation between these two states of about 150 cm-1 [5]. The displacement of the iron atom from the heme plane in deoxyHb is rather uncertain, a value of 0.6 Å has been commonly admitted [6] but another estimate places the iron only 0.2-0.3 Å out of the plane [7].

Ideally one may hope that the study of synthetic models will help to answer the questions regarding the structure and the electronic properties of the heme. Our understanding of the iron-dioxygen linkage in oxyHb has certainly been improved by the studies of a number of synthetic iron(II) oxygen carriers since it has been shown unambiguously that the dioxygen ligand is bound end-on (see for instance reference [8]). The design of a high-spin five-coordinate Fe(II) porphyrin (the so-called picket fence porphyrin) and of its dioxygen adduct has allowed to study the structural changes occuring upon oxygenation [9]. However these studies are not completely free of uncertainties such as the following ones :

- the O-O separation of 1.16 Å in the dioxygen complex of the picket fence porphyrin is unrealistically short when compared to the value of 1.21 Å for molecular oxygen [10]. A kinked structure such as 3 (with

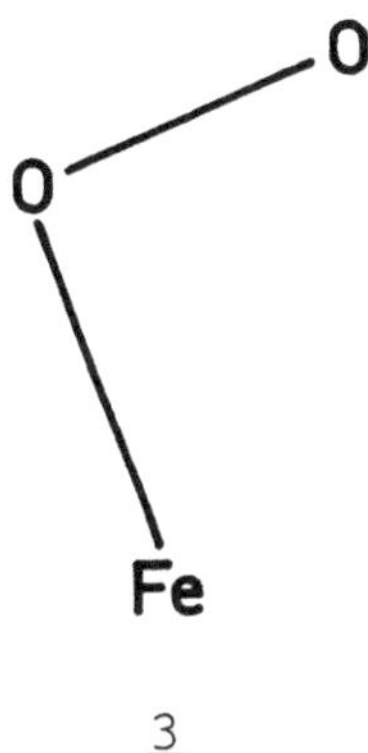

3

the Fe-O bond off the axis perpendicular to the porphyrin plane) could not be ruled out. Since the superoxide anion O_2^- has a bond length of 1.34 Å [11], this short O-O separation is at variance with the Fe(III)-O_2^- formulation proposed by the same authors on the basis of the infrared spectrum [12] (the corresponding $\nu(O_2)$ frequency at 1159 cm-1 is extremely close to the values for the free superoxide ion, 1145 cm-1, and for a number of Co(II)-Schiff-base dioxygen adducts formulated as Co(III)-O_2^- [13]).

- although this dioxygen adduct is diamagnetic, another dioxygen complex of the picket-fence porphyrin was reported to be paramagnetic[14]. However this observation was found latter to be artifactual [15].
- the deoxyHb models all have a 2-MeIm (Im=imidazole, Me=methyl) ligand. The methyl group probably experiences a repulsive interaction with the porphyrin ring :

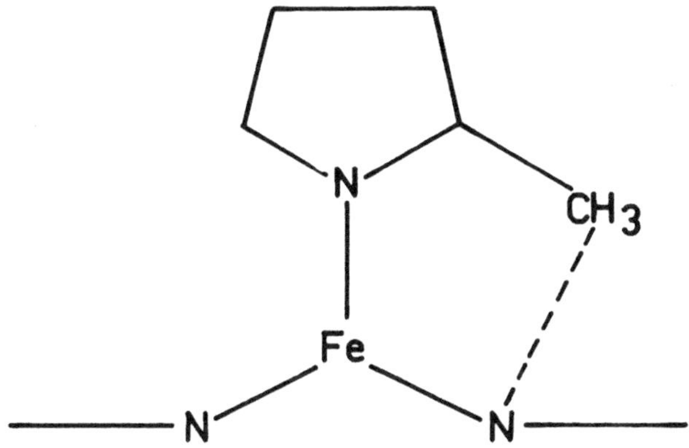

and this interaction may increase the out-of-plane displacement of the iron from its intrinsic, unstressed value [8].

In principle, quantum mechanical studies are ideally suited to answer the many questions yet unsolved regarding the structure and electronic properties of the heme :

- what are the electronic configuration and the geometry of the iron-dioxygen unit in the dioxygen complex ?

- what is the relationship between the electronic structure of the heme and its spectroscopic properties such as the O-O stretching frequency and the quadrupole splitting in the ^{57}Fe Mössbauer spectra (the quadrupole splitting is related to the electric field gradient at the iron nucleus [16]) ?

- what is the out-of-plane displacement of the iron in a five-coordinated, high-spin, iron(II)porphyrin ? Does the globin influence the out-of-plane displacement of iron in deoxyHb ?

The theoretical study of large biological molecules such as the iron porphyrins has been restricted for many years to the semi-empirical methods of quantum chemistry [17]. There has been recently a revival of interest for the application of these semi-empirical methods to heme studies [18-21]. In the same time the considerable developments of the methodology together with the new computational facilities have made it possible to treat at the _ab initio_ level these relatively large systems. We present here the results of _ab initio_ LCAO-MO-SCF calculations on heme models (ab initio calculations have also been reported by Goddard on less realistic models of the heme [22,23]).

GENERAL ASPECTS OF THE CALCULATIONS

Ab initio LCAO-MO-SCF calculations including all the electrons were carried out for the model systems $FePNH_3$ (representing the heme in deoxyHb, P = porphine dianion) (Fig. 4), $FePO_2NH_3$ (for the heme in oxyHb) (Fig. 5) and FeP. Limitations in the size of the program and in computing costs forced us to mimic the imidazole ligand of the heme through an ammonia molecule. However one calculation was also carried out for the system $FePO_2Im$ and a comparison of the results with those of the model calculation indicate that the approximation used is a relatively safe one [24]. The calculations were not carried out at the same level of accuracy for the deoxyheme and the oxyheme models. For the oxyheme model, the search on the potential energy surface of the iron-dioxygen unit required a relatively large number of calculations and the use of a minimal basis set except for the d-shell which is split (one calculation with a basis set of double-zeta quality for the valence shells has been carried out in order to check the results of the minimal basis set calculations [25]). For the deoxyheme model, the calculation was intended to yield a relatively accurate value of the out-of-plane displacement of the iron atom and this required the use of a basis set of double-zeta quality at the level of the valence shells (but minimal for the inner shells). Finally, the calculations on the FeP system in the intermediate

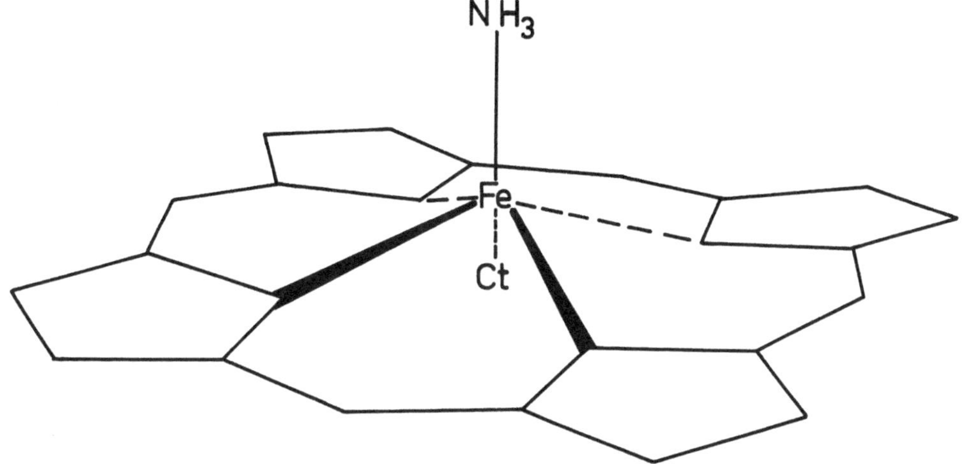

Fig. 4 The $FePNH_3$ model.

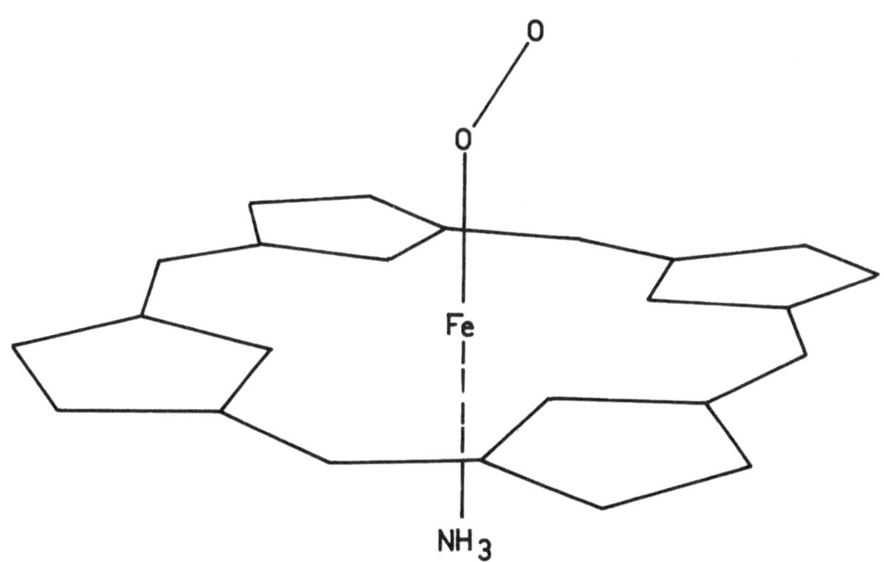

Fig. 5 The $FePO_2NH_3$ model for the bent structure.

spin (S=1) were intended as a test of the accuracy of the wavefunctions by calculating the value of the quadrupole splitting in the Mössbauer spectra. In order to assess the effect of the basis set, two basis sets were used, one of double-zeta quality at the level of the valence shells (again minimal for the inner shells) and the other one only slightly better in terms of the number of contracted functions, but built from a larger Gaussian basis set. Table I summarizes these various basis sets.

Since these calculations are relatively expensive in terms of computer time and since they are presently restricted to the SCF level, a limited use of experimental data is unavoidable both in order to minimize the number of calculations and to avoid the pitfalls of the correlation error. For this reason we have considered only electronic states corresponding to the "experimental" spin value (namely S=2 for $FePNH_3$, S=0 for $FePO_2NH_3$ and S=1 for FeP)(however calculations were also carried out for a number of triplet states S=1 of the oxyheme model). Somewhat idealized geometries were used throughout the calculations, assumed values (taken from experimental structures for related systems) were given to the bond lengths and bond angles which were not optimized (for the details of the geometries, see References [24-26]).

The calculations were carried out with the system of programs Asterix [27]. These programs have been developped mostly for the calculation of relatively large systems involving transition metals and high efficiency has been achieved through a number of features. For instance the evaluation of the two-electron integrals over gaussian functions of high angular momentum relies on the block techniques [28] together with the full exploitation of the symmetry operations pertaining to non-degenerate point groups [29,30]. The essential features of the SCF programs have been described previously [31].

THE OXYHEME MODEL

Electronic configuration of the iron-dioxygen unit

Fig. 6 shows the sequence of electronic states based on the SCF energies calculated for various electronic configurations of $FePO_2NH_3$ with the bent structure [24,25]. The electronic configurations considered in the SCF calculations fall within two classes : i) the closed-shell configurations $d^6\pi_g^2$ where d^6 stands for $d_{xy}^2 d_{xz}^2 d_{yz}^2$ (these are the t_{2g} orbitals corresponding to the local O_h pseudo-symmetry of the iron atom) and π_g stands either for π_g^a (4) or π_g^b (5) ; these configurations give rise to singlet states ; ii) the open-shell configurations $d^5(\pi_g^a)^2(\pi_g^b)^1$ corresponding to the promotion of one electron from one of the above 3d orbitals to the π_g^b orbital ; these configurations have two unpaired electrons, one on the iron and the other on dioxygen, and they give rise to triplet and singlet states. The lowest SCF energy corresponds to a paramagnetic triplet state $^3A'$ with a formal configuration Fe(III)-O_2^- of the type $d^5\pi_g^3$. The singlet state $^1A'$ corresponding to the configuration $d^6(\pi_g^a)^2$ of the type Fe(II)-O_2, originally proposed by

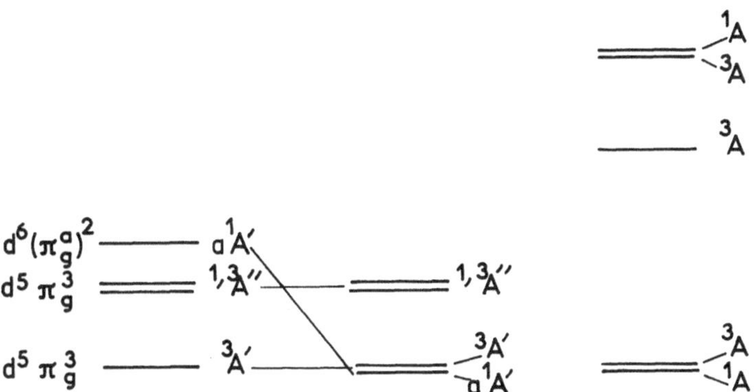

Fig. 6 The energies of the electronic states of the oxyheme model. Left, SCF energies for $FePO_2NH_3$ based on the *ab initio* calculations. Middle, a qualitative diagram obtained by adding an empirical correction for the correlation energy to the SCF values. Right, the INDO-SCF-CI energies of Ref. [21] for the model system $FePO_2Im$.

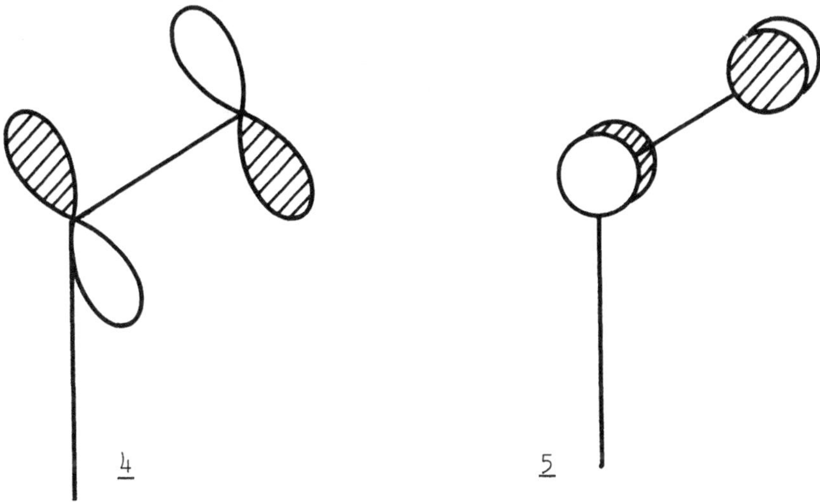

Table I. Basis sets and number of electrons for the systems considered.

	Gaussian basis set	Contracted basis set	Nb. of gaussian functions	Nb. of contracted functions	Nb. of electrons
FeP	(11,7,5/8,4/4)	[5,4,2/3,2/2]	590	269	186
	(13,9,6/9,5/4)	[5,4,3/3,2/2]	700	275	186
$FePNH_3$	(11,7,5/8,4/4)	[5,4,2/3,2/2]	622	284	196
$FePO_2NH_3$	(10,6,4/7,3/3)	[4,3,2/2,1/1]	529	175	212

Pauling as the ground state configuration [2], is 0.031 a.u. (0.8 eV) higher in energy (the state $^1A'$ corresponding to the configuration $d^6(\pi_g^b)^2$ is higher). However, we have to take into account that the correlation error bias these SCF calculations in favor of the open-shell configuration since it has one electron pair less than the closed-shell configuration. Since the difference in correlation energy between the closed-shell and open-shell configurations would be of the order of a pair correlation energy, about 0.03 a.u. (and probably larger due to the near-degeneracy of the two orbitals of dioxygen π_g^a and π_g^b [25]), we assigned as the ground state the singlet state $^1A'$ with the closed-shell configuration $d^6\pi_g^2$ and we predicted that a triplet state $^3A'$ should be very close in energy to the ground state singlet [24] (Fig. 6). This prediction has been corroborated by further experimental and theoretical work. The interpretations given by Cerdonio et al of their magnetic measurements [5] and by Koster of the iron Kβ fluorescence emission spectrum of oxymyoglobin [32] postulate the existence of a triplet state extremely close to the diamagnetic ground state. The same feature is found in the diagram of electronic states obtained by Loew et al through semi-empirical INDO-SCF-CI calculations for a model $FePO_2Im$ [21] (Fig. 6). For the perpendicular structure, the lowest energy closed-shell configuration turns out to be $d^6(\pi_g^b)^2$ (π_g^b as shown in 6) rather than $d^6(\pi_g^a)^2$ [24].

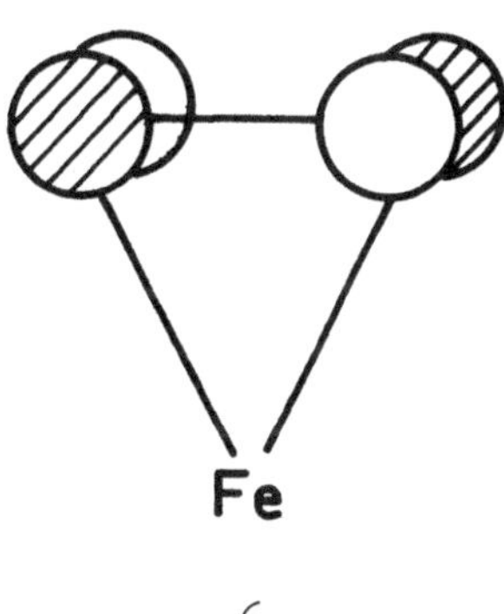

6

This is a consequence of the fact that the metal-dioxygen interaction 7 in the perpendicular structure destabilize π_g^a (relatively to π_g^b which shows practically no interaction with the iron 3d orbitals) whereas in the bent structure π_g^b is more destabilized by interaction 8 than π_g^a is destabilized through 9.

Bent, kinked and perpendicular structures

The perpendicular structure is found to be less stable than the bent one by 55 kcal/mole. This is in agreement with the observation of a bent iron-dioxygen unit in the structures of the dioxygen complex of the picket fence porphyrin [8], of oxymyoglobin [33] and of oxyerythrocruo-

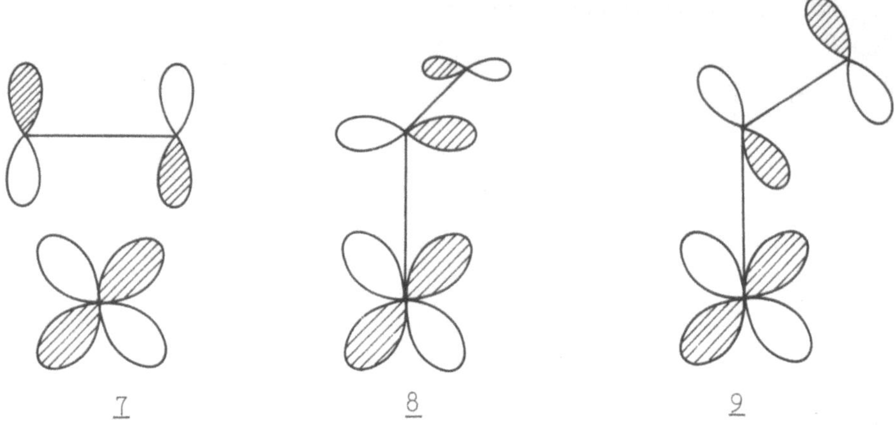

rin [34]. The above value is large enough to rule out the possibility of a perpendicular structure in the oxyheme (further theoretical refinements such as basis set extension or geometry optimization may change this value slightly but certainly not enough to make the perpendicular structure more stable than the bent one). The reason for this large destabilization of the perpendicular structure (or at least the main factor) is a rather simple one : in this structure, the distance between one oxygen atom and the nearest pyrrole nitrogen is 2.28 Å, far less than the sum of 2.9 Å for the corresponding van der Waals radii [35]. This certainly gives rise to a repulsive interaction which could be reduced only by increasing the iron-dioxygen distance (but this could certainly weaken the bond) or by pulling the iron atom out of the porphyrin plane (but the trans ligand will certainly oppose this displacement of the iron atom). Perpendicular coordination of the dioxygen ligand has been unequivocally established for two metalloporphyrins. One is the peroxotitaniumoctaethylporphyrin, with the titanium atom five-coordinate and displaced from the porphyrin plane by 0.62 Å [36] (with a calculated contact between oxygen and nitrogen of 2.6 Å). The other one is a trans diperoxomolybdenum(VI)porphyrin [37], with the Mo atom in the porphyrin plane and again with a short contact of 2.30 Å between one oxygen atom and the nearest nitrogen. However, in these two systems the repulsive interaction is balanced by orbital factors, namely a stabilizing interaction 10 with a large overlap term (this is tantamount to a δ bond) [25,38] (electronic factors may also destabilize the perpendicular structure in $FePO_2NH_3$ relative to the bent structure [24]).

The calculation for a kinked structure such as 3 (with the oxygen molecule displaced such that the iron atom nearly bisects the projection of the O-O bond onto the heme plane) shows that it is about 6 kcal/mole higher in energy than the bent structure 1 with the Fe-O axis perpendicular to the porphyrin plane. Thus our calculations do not support the results of the extended Hückel calculations [18] which predict that the ground state geometry of oxyheme corresponds to this kinked structure. An off-axis displacement of the bonded oxygen atom has been conside-

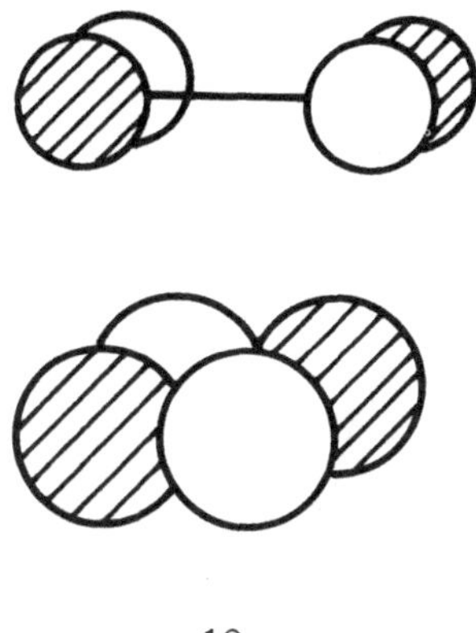

10

red as likely in the structure of the dioxygen complex of the picket fence porphyrin, mostly in order to account for the very short O-O separations of 1.15 and 1.17 Å [10] (a small displacement of the bonded oxygen atom would lead to an increased O-O separation as shown in 11).

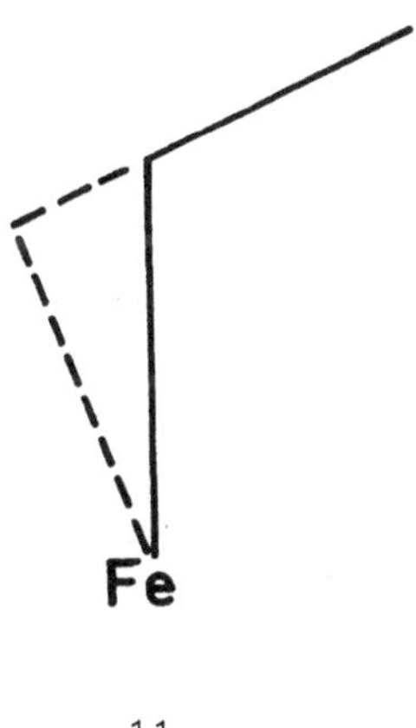

11

One cannot rule out the possibility that the bent structure 1 represents the equilibrium geometry for the unsubstituted ferroporphyrin whereas the kinked structure would become more stable for the highly substituted, sterically encumbered, picket fence porphyrin. This hypothesis would be supported by the observation, for the picket fence complex, of an abnormally short separation of 2.68 Å between the terminal oxygen atom and one methyl carbon of the pivalamide picket [10], well below the sum (3.4 Å) of the van der Waals radii for an oxygen atom and a methyl group. This short contact may induce a deformation of the FeO_2 unit towards a kinked structure (Fig. 7). The small energy difference found

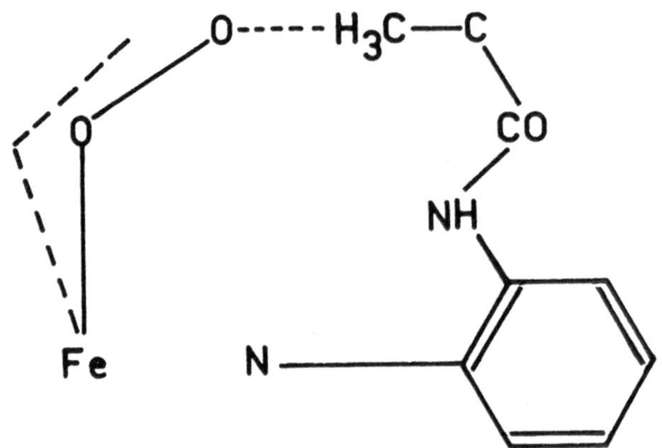

Fig. 7 The steric interaction between the dioxygen terminal atom and one methyl group of the pivalamide picket in the dioxygen complex of the picket fence porphyrin (according to the structural data of Ref.[10]) and the partial releave of the interaction through a deformation to a kinked structure.

between the two structures 1 and 3 indicates that the potential energy surface is rather flat in this region and that this type of deformation should be relatively easy (a soft potential energy surface has also been found for the bending or tilting of the FeCO unit in a model of the carboxyhemoglobin [25]).

The O-O bond length, the FeOO bond angle and the O-O stretching frequency

Optimization of the FeOO angle in the bent structure 1 yields a value of 128° in good agreement with the experimental angle of 131° [10]. Optimization of the O-O bond length produces a value of 1.30 Å which is certainly overestimated since the optimized bond length for the ground state of O_2, with the same minimal basis set, is also 1.30 Å vs. an experimental value of 1.21 Å (bond lengths calculated with a minimal basis set are usually too long [39]). Given the error due to the use of a minimal basis set, we believe that the O-O separation in deoxyheme falls probably within the range 1.22 - 1.26 Å. Hoard considers that it lies probably in the range between the superoxide and peroxide values of about 1.28 and 1.48 Å respectively [40]. This is a consequence of the idea that the electronic structure of the dioxygen ligand in oxyheme is close to that of the superoxide ion. Support for this analogy is usually provided by the similarity in the O-O stretching frequencies of the dioxygen complex of the picket fence porphyrin (1159 cm-1) and of the superoxide anion (1145 cm-1) (compared to the value of 1556 cm-1 for molecular dioxygen). In fact, the calculation of the stretching frequencies for dioxygen and for the oxyheme model reproduces extremely well the decrease which is observed in the experimental frequencies upon complexation of molecular dioxygen [41] (Table II), although the wavefunction of

Table II. Calculated O-O stretching frequencies versus experimental values (in cm-1).

	Calculated O-O frequency	Experimental O-O frequency
O_2	1741	1556
$FePO_2NH_3$	1326	1160 [a]

[a] In Fe(TpivPP)(NMeIm)O_2

the oxyheme model has an electronic configuration of the type Fe(II)-O_2 rather than Fe(III)-O_2^- (the calculated stretching frequencies are too large but this is a general feature of the minimal basis set calculations [42]). This decrease in the O-O frequency is a consequence of the charge reorganization brought about in the dioxygen ligand upon complexation. A synergic process involves a charge transfer from the dioxygen orbitals $3\sigma_g$ and π_g^a to the $3d_{z^2}$ orbital of iron, while π backbonding populates the π_g^b orbital of dioxygen at the expense of one metal d_π orbital (Fig. 8) (for the sake of simplicity we refer to a dioxygen ligand in a $^1\Delta$ state). Both charge transfers weaken the O-O bond since the $3\sigma_g$ orbital is bonding while the π_g^b orbital is antibonding.

Rotational isomerism

The stereochemistry of the oxyheme model is characterized further by the orientation of the plane of the FeO_2 unit with respect to the porphyrin, in other words the O-O bond may project either along the Fe-N bonds (eclipsed structure, 12 represents the Newman projection along the Fe-O axis) or along the bisector of the angle ∠NFeN (staggered structure 13). For the sake of economy the calculation of the relative stability of these two rotamers was carried out on a five-coordinate model $FePO_2$. The most stable conformation corresponds to the staggered structure 13 while the eclipsed structure 12 is about 6 kcal/mole higher. This is in agreement with the crystal structure of the dioxygen complex of the picket fence porphyrin, where the FeOO plane was found to bisect the ∠NFeN angle [10]. An eclipsed arrangement of the FeOO plane relative to N-Fe-N has been found in oxymyoglobin [33] and explained on the basis of steric hindrance between the dioxygen ligand and nearby residues of the globin. The experimental report of both a staggered and an eclipsed conformation is in agreement with the small energy difference produced by the calculations. The iron-dioxygen interactions favor the eclipsed conformation since it is weakly stabilized by the interac-

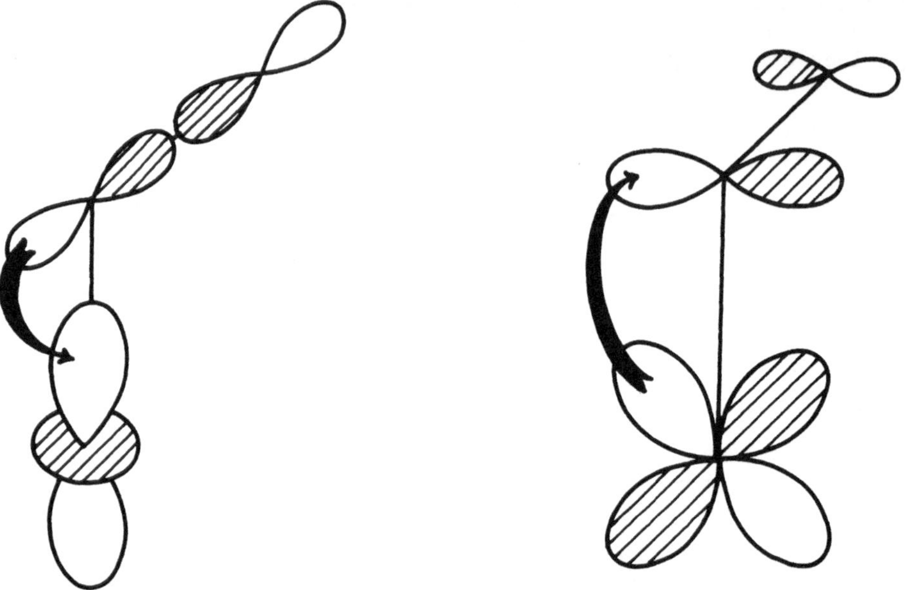

Fig. 8 The synergic charge transfers between iron and dioxygen.

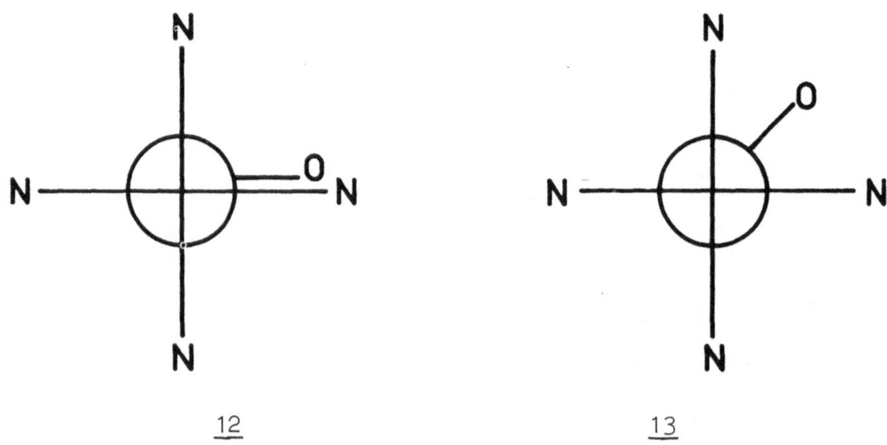

tions 14 (π_g^a-$d_{x^2-y^2}$) and 15 (π_g^b-d_{xy}) whereas the staggered conformation is somewhat destabilized by a four-electron destabilizing interaction 16 (π_g^a-d_{xy}) (Fig. 9) (only the interactions between the dioxygen π_g orbitals and the metal $3d_{xy}$ and $3d_{x^2-y^2}$ orbitals have been included in the diagram of Fig. 9, since the interactions with the metal $3d_{xz}$, $3d_{yz}$ and $3d_{z^2}$ orbitals will be the same for the two conformations). We trace the destabilization of the eclipsed conformation to the steric interaction between the terminal oxygen atom and the nearest porphinato nitro-

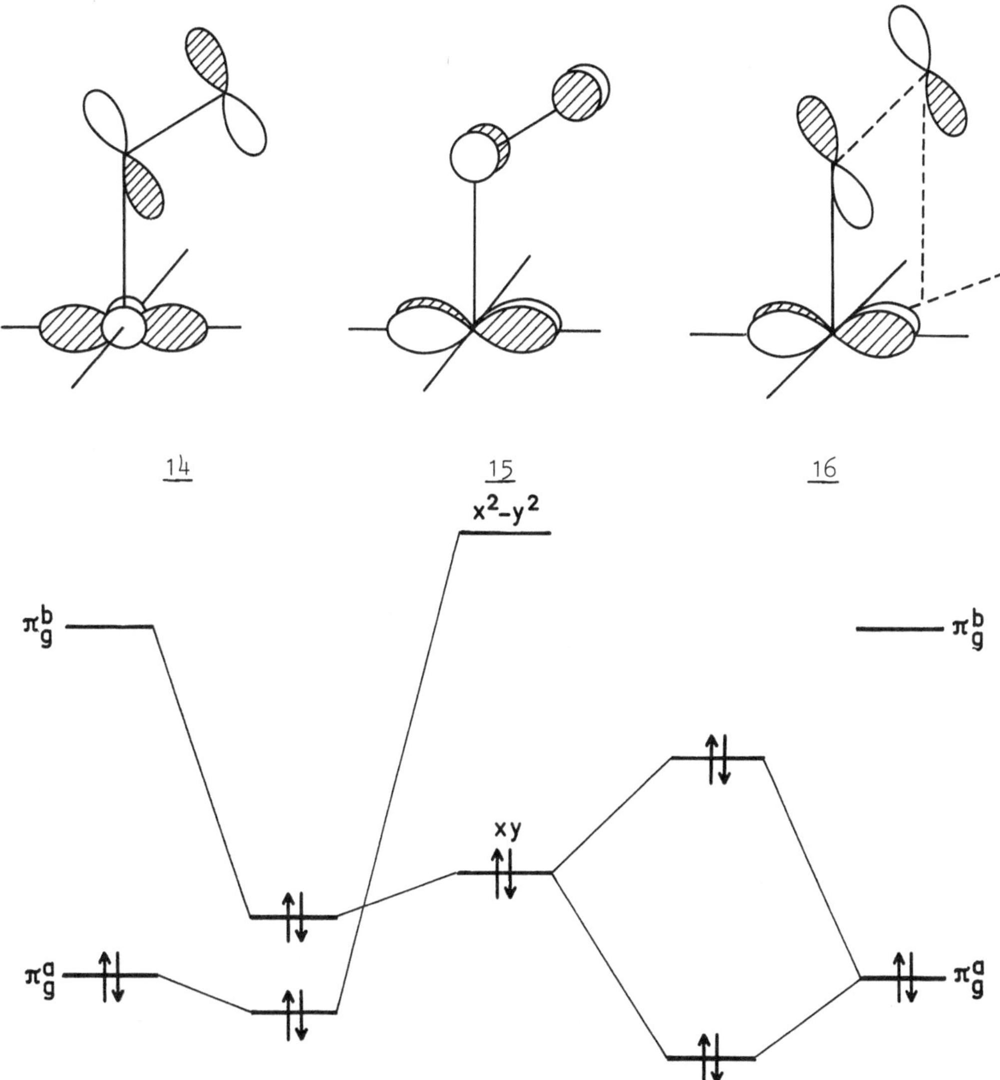

Fig. 9 The interaction diagram between the metal $3d_{xy}$ and $3d_{x^2-y^2}$ orbitals and the dioxygen π_g orbitals for the two conformations, eclipsed (left) and staggered (right) of the oxyheme model.

gen atom, since the calculated contact between these two atoms is 2.86 Å for the eclipsed structure, thus less than the sum of their van der Waals radii (2.90 Å) but is slightly greater for the staggered structure (3.03 Å) [25]. Eclipsed conformations have been reported experimentally for the peroxo complexes of the Ti(IV) and Mo(VI) porphyrins[36,37], they have been rationalized on the basis of the metal-ligand interactions [38,43].

Hydrogen bonding to the distal imidazole

A problem which has been pending for a long time concerned the stabilization of oxyheme through hydrogen bonding between the dioxygen ligand and the distal imidazole, as shown in 17. Although originally

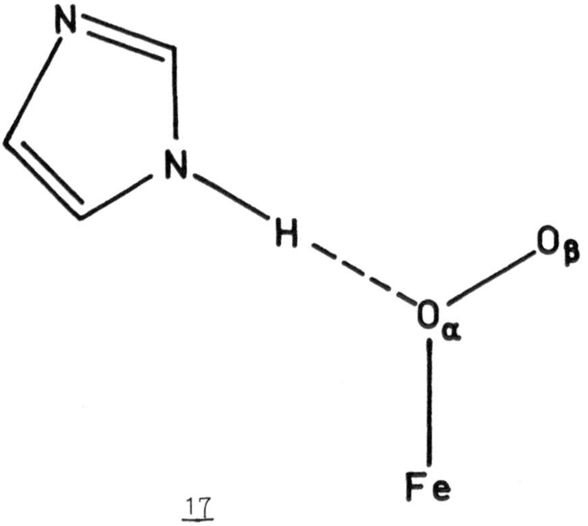

17

Pauling suggested hydrogen bonding through the terminal oxygen atom O_β [44], the crystallographic data [45] show that the O_α atom is at about 2.5 Å from the $N_{\varepsilon 2}$ atom of the distal histidine, a situation where the contact is close enough for hydrogen bonding. A calculation carried out on a model system $FePO_2(NH_3)_2$ with a N_ε-O_α distance of 2.5 Å yields a strongly repulsive interaction [24] (with a destabilization energy of 24 kcal/mole) and rules out such hydrogen bonding (the same conclusion has been reached on the basis that the picket fence porphyrin, which lacks the distal histidine, reproduces well the thermodynamic constants for the binding of O_2 found in the biological systems [46]). The strongly repulsive interaction is certainly a consequence of the relatively short N-O distance, as shown by the potential energy curve for the model system NH_3-O_2 [41] (Fig. 10). The lack of hydrogen bonding both in our model system and in oxyHb results probably from the neutral character of the dioxygen ligand, since the model system NH_3-O_2^- shows rather strong hydrogen bonding to the superoxide anion [41] (Fig. 10).

THE DEOXYHEME MODEL

Electronic configuration of iron in deoxyheme

In deoxyheme the iron atom is five-coordinate, the non-bonding orbitals d_{xy}, d_{xz} and d_{yz} are nearly degenerate, the antibonding orbi-

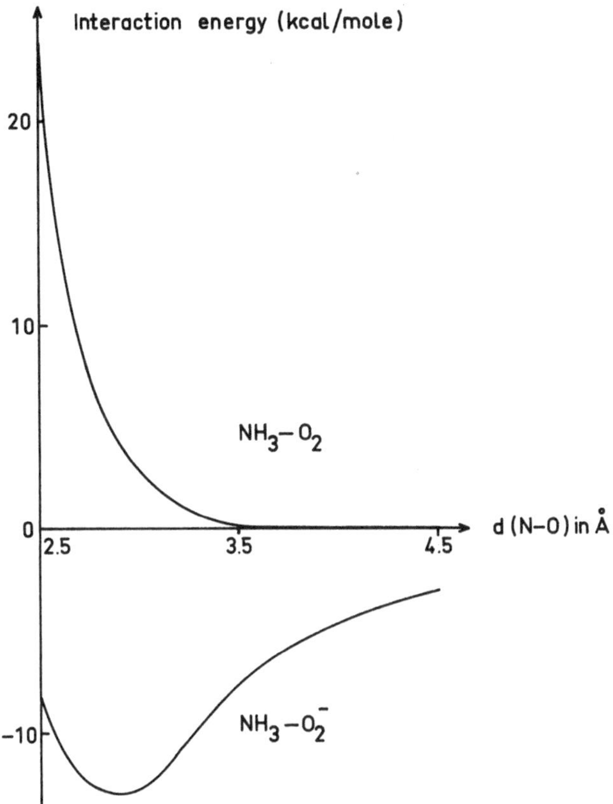

Fig. 10 The potential energy curves for the interaction between an ammonia molecule and an oxygen molecule or a superoxide anion (one proton is located between the nitrogen and oxygen atoms).

tal $d_{x^2-y^2}$ is strongly destabilized through the interactions with the porphyrin nitrogens while the orbital d_{z^2} is less destabilized through the interaction with the axial ligand (the four nitrogen atoms of the pyrrole rings are along the x and y axis). Since the iron atom of the deoxyheme is high-spin (S ≐ 2), the electronic configurations which are the best candidates for the ground state are the ones of Fig. 11. The ground state of the high-spin, five-coordinate model $FePNH_3$ corresponds, according to the energy values of Table III, to the configurations $^5A''$ $(xy)^1(xz)^1(yz)^2(z^2)^1(x^2-y^2)^1$ and $^5A'$ $(xy)^1(xz)^2(yz)^1(z^2)^1$ $(x^2-y^2)^1$ which are practically degenerate. They are also the ground state configurations in the extended Hückel calculations [20] and in the GVB-CI model calculations [23]. The strongest argument for assigning the ground state configuration is provided by the asymmetry parameter in the Mössbauer spectra, with an experimental value in the range 0.5-0.7 for deoxyHb and deoxyMb [47,48] and the theoretical value equal to zero for the configuration $(xy)^2(xz)^1(yz)^1(z^2)^1(x^2-y^2)^1$ but different from zero

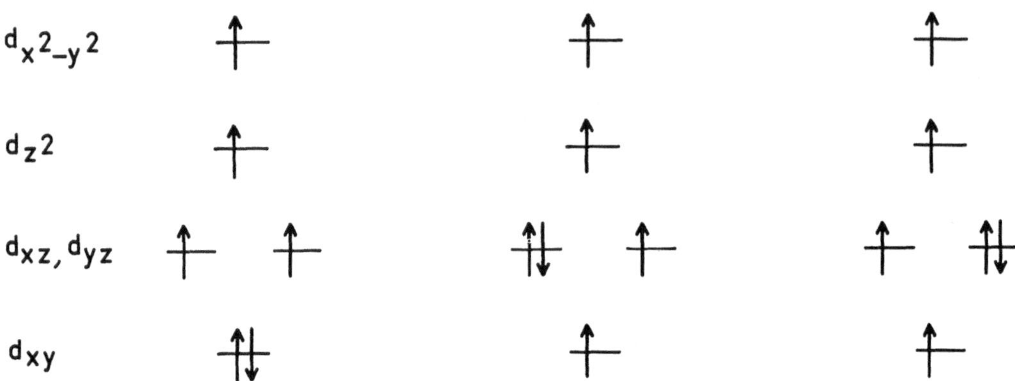

Fig. 11 The electronic configurations of lowest energy for the deoxyheme model.

Table III. Total SCF energies (in a.u.) for three electronic configurations of $FePNH_3$ (S=2) with the iron atom in the porphyrin plane.

Configuration	Energy
$(xy)^1(xz)^1(yz)^2(z^2)^1(x^2-y^2)^1$	-2297.2574
$(xy)^1(xz)^2(yz)^1(z^2)^1(x^2-y^2)^1$	-2297.2573
$(xy)^2(xz)^1(yz)^1(z^2)^1(x^2-y^2)^1$	-2297.2551

for the other two configurations [26], thus ruling out the former configuration.

The out-of-plane displacement of the iron atom in the deoxyheme model

Fig. 12 shows the potential energy curve for our deoxyheme model as a function of the out-of-plane displacement of the iron for the ground state configuration. The curve has a minimum for a displacement of 0.32 Å. This value is slightly smaller than the experimental values of 0.42 and 0.40 Å reported for the five-coordinate porphyrins with a 2-MeIm ligand [9]. The steric requirements of the 2-MeIm ligand are probably responsible for the increased displacement of about 0.1 Å. A similar trend is found in the dioxygen complexes of the picket fence porphyrin, with the iron atom displaced 0.086 Å out of the plane towards the imidazole ligand in the case of a 2-MeIm ligand, in contrast to a displacement of 0.03 Å towards the O_2 ligand for a 1-MeIm imidazole [9]. The adjustement for the steric hindrance of the 2-MeIm group in these

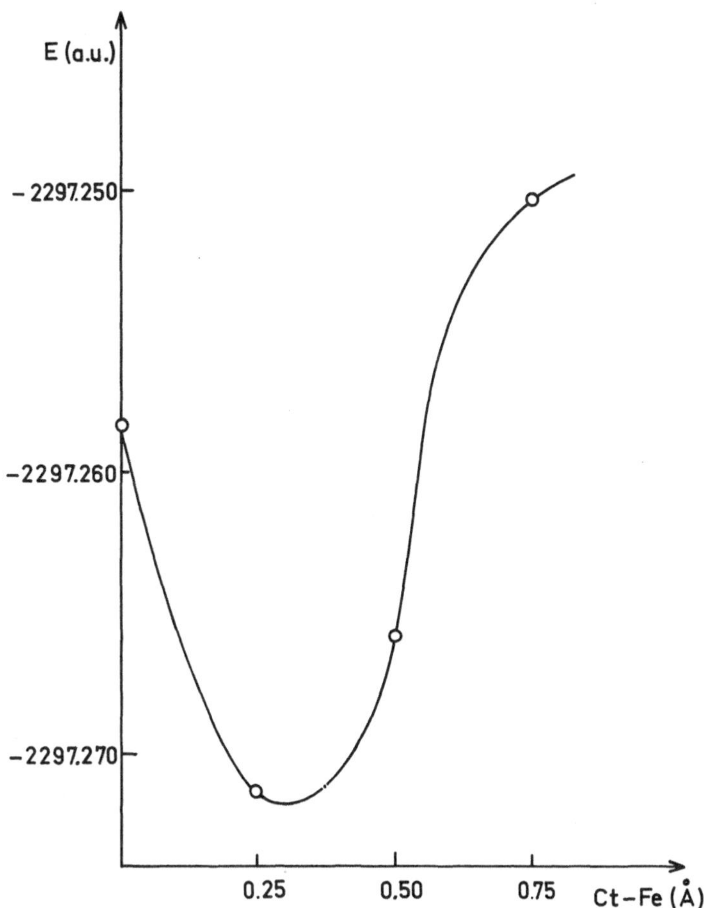

Fig. 12 The potential energy of $FePNH_3$ as a function of the out-of-plane displacement of the iron.

molecules is made through a displacement of 0.11-0.12 Å of the iron atom, a value close to the difference of 0.08-0.10 Å between the computed displacement in our deoxyheme model and the experimental values for a 2-MeIm ligand. The calculated displacement of 0.32 Å is also very close to the experimental value of 0.28 Å, based on a X-ray crystal structure for a polymeric picket fence porphyrin with the iron atom of one porphyrin unit coordinated to the pivalamid oxygen atom of another unit, this system being devoid of any short contact between the axial ligand and the porphyrin ring [49]. This comparison with the experimental data gives some confidence in the calculated value of the out-of-plane displacement. Rather similar values have been reported in two other theoretical studies (0.2 Å in an extended Hückel calculation for the model system FePIm [20] and 0.28 Å in a GVB calculation on the model $Fe(NH_2)_4(NH_3)$ [23]).

The calculated value of 0.32 Å for the out-of-plane displacement of the iron atom is significantly smaller than the accepted value of 0.6 Å for the extent of non-planarity in human deoxyhemoglobin [6]. If one accepts the latter value, this means that the protein can keep the iron atom in a tense conformation, by preventing it from moving to its equilibrium position. The energy required to shift the out-of-plane displacement of iron from an equilibrium value of 0.3 Å to the value of 0.6 Å would be of the order of 10 kcal/mole according to Fig. 12, this is somewhat larger than the energy of cooperativity (of the order of 3 kcal/mole per heme [1]), but one cannot rule out the possibility that either the displacement of iron in deoxyhemoglobin has been overestimated and/or that our calculation underestimates the displacement of iron in an unconstrained system. A number of proposals have been made which either support or question the tension theory. For instance Perutz proposed that "the displacement of the iron atom from the plane of the porphyrin in tetrameric deoxyhemoglobins in the T state should be greater than in monomeric ones which are not constrained and have higher oxygen affinities" [50]. Unfortunately, there are not many experimental measurements of the non-planarity of the iron for high-affinity forms such as the isolated α and β subunits of deoxyhemoglobin or for abnormal deoxyhemoglobins which cristallize in the R form [50,51]. A number of arguments have also been given in support of a lack of strain on the heme group in the deoxy quaternary structure, based on empirical energy calculation [52], Mössbauer experiments [47], EXAFS measurements [53] and resonance Raman spectra [54]. However some of the conclusions derived from these experiments have been criticized [20]. The extreme confusion in this area is exemplified by the fact that an EXAFS study of deoxyhemoglobin has yielded an out-of-plane displacement of the iron in the range 0.2-0.3 Å [7]. Should this result be corroborated, then there would be no tension at the iron in the heme.

The origin of the out-of-plane displacement of iron

A question which is frequently addressed is the origin of the out-of-plane displacement of the iron [55]. Traditionally, one considers that the bonding radius of the high-spin iron is too large to fit into the porphyrin hole [56]. An equivalent explanation emphasizes the presence of an electron in the antibonding orbital $d_{x^2-y^2}$, since pulling the iron out of the porphyrin plane will reduce the antibonding character of this orbital [57]. According to Olafson and Goddard, the origin of the out-of-plane displacement for the high-spin, five-coordinated iron is rather to be found in the nonbonded repulsive interactions between the electron pairs of the fifth ligand and of the pyrrole nitrogens, since a four-coordinated, high-spin iron of a model system $Fe(NH_2)_4$ is found to lie into the plane of the four NH_2 groups [23].

Calculation of the energy for the high-spin (S = 2) four-coordinate system FeP shows that the equilibrium geometry corresponds indeed to the iron atom in the porphyrin plane [26]. This supports the proposal of Goddard that the out-of-plane displacement of iron for the five-coordinated porphyrins is not due to the size of the high-spin iron(II) but rather to the non-bonded interactions between the fifth ligand and the pyrrole nitrogens of porphyrin.

A SIMPLE SYSTEM : THE Fe(II) PORPHYRIN

One of the simplest iron(II) porphyrins in terms of coordination is probably the four-coordinated α,β,γ,δ-tetraphenylporphinatoiron(II), FeTPP [58]. In this molecule the iron atom is four-coordinated with an intermediate spin value (S = 1) and is found at the center of the molecule [58]. Yet the electronic ground state of this molecule has bot been established conclusively. Three electronic configurations appear compatible with the spin value on simple orbital energy grounds :

(a) $(xy)^1(xz)^2(yz)^2(z^2)^1$ $^3B_{1g}$
(b) $(xy)^2(xz)^1(yz)^1(z^2)^2$ $^3A_{2g}$
(c) $(xy)^2(xz)^{1.5}(yz)^{1.5}(z^2)^1$ ^3E_g

(we label these states according to D_{4h} symmetry, with the orbitals $3d_{xz}$ and $3d_{yz}$ degenerate, although the FeTPP molecule shows only S_4 symmetry due to extensive ruffling of the porphinato core). Analysis of the proton NMR spectra indicates large π contact shifts, thus requiring unpaired spins in d_{xz} and d_{yz} and suggests that the orbital ground state is non degenerate, which leaves configuration (b) as the only candidate for the ground state configuration [59]. However, when the electric field gradient is calculated as a sum of atomic 3d contributions [60], all three configurations are predicted to give a negative quadrupole interaction in the ^{57}Fe Mössbauer spectrum [58] in contrast to the positive value observed experimentally [61] (Table IV). Support for configuration (b) has also been gained from the fact that it fits the magnetic data [58]. An extended Hückel calculation predicts an ^3E_g ground state (configuration (c)) [62], and the resonance Raman spectra of ferrous octaethylporphyrin (FeOEP), another four-coordinated iron(II)porphyrin, has also been interpreted in terms of an ^3E_g ground state [63].

SCF energy values (Table IV) point to the state $^3A_{2g}$ being lower in energy than the state $^3B_{1g}$. Furthermore the value calculated for the quadrupole splitting of the $^3B_{1g}$ state, from the SCF wavefunctions, is large and negative, in contrast to the experimental value which is positive. This probably rules out the $^3B_{1g}$ state as the ground state and leaves configurations (b) and (c) as the best candidates for the ground state. No direct comparison could be made in terms of energy values since our open-shell SCF program was restricted to the half closed-shell case and could not handle the ^3E_g state. However one will notice that the calculated quadrupole splitting is negative but small for configuration (b) while it is large and negative for configuration (c). Thus we select configuration (b) as the ground state configuration.

The quadrupole splitting depends on the electric field gradient at the iron nucleus, which represents an expectation value of a one-electron operator. It is generally considered that the Hartree-Fock value of the one-eleetron operators should be accurate to second order, a consequence of Brillouin's theorem [64]. Yet the calculated quadrupole splitting is negative in contraet to the positive value reported experi-

Table IV. SCF energies and quadrupole splittings for several electronic configurations of FeP (S = 1).

	Electronic configuration		SCF energy (a.u.)		Quadrupole splitting [a]		
					[b]	SCF values (mm/s)	
			Basis set I	Basis set II		Basis set I	Basis set II
(a)	$(xy)^1(xz)^2(yz)^2(z^2)^1$	$^3B_{1g}$	-2241.0690	-2240.8377	- 8/7	-4.8	- 5.8
(b)	$(xy)^2(xz)^1(yz)^1(z^2)^2$	$^3A_{2g}$	-2241.0951	-2240.8660	- 4/7	-0.3	- 0.7
(c)	$(xy)^2(xz)^{1.5}(yz)^{1.5}(z^2)^1$	^3E_g	-	-	- 2/7	-2.8	- 3.6
(d)	$(xy)^2(xz)^1(yz)^1(x^2-y^2)^2$	$^3A_{2g}$	-	-2240.5907	+12/7		+19.3
(e)	$(xz)^1(yz)^1(z^2)^2(x^2-y^2)^2$	$^3A_{2g}$	-	-2240.7050	- 4/7		- 0.1

[a] Experimental value 1.51 mm/s [61].

[b] Atomic 3d contributions only, expressed in units of $\frac{1}{2}e^2Q < r^{-3} >$ [60].

mentally. One may argue that the calculations have not reached the Hartree-Fock limit for FeP, however one sees no improvement upon going to a larger basis set (Table IV). Remember that the electric field gradient value is mostly dependent on the d orbitals and that the larger basis set has three contracted functions of the d type (that the total energy is lower for the small basis set is merely a consequence of the fact that the 3s and 3p shells were better described with the small basis set). The situation is reminiscent of the one for the dipole moment of CO, with the Hartree-Fock value and the experimental value of opposite sign [65]. The quadrupole splitting of FeP is small in absolute magnitude (the same was true for the dipole moment of CO). Besides configuration (b), there are two other $3d^6$ configurations of $^3A_{2g}$ symmetry (d) and (e) which can mix with (b) in the ground state wavefunction (corresponding to the promotion of two electrons to the $d_{x^2-y^2}$ orbital). One will notice that configuration (d) is characterized by an extremely large, positive value of the quadrupole splitting. Thus a small admixing of configuration (d) in the ground state wavefunction could change aprreciably the calculated value of the quadrupole splitting, including the sign. Furthermore the SCF calculations have been carried out with the assumption of a planar geometry for the porphyrin core, although there is extensive ruffling of the porphinato core in the FeTPP molecule. Preliminary calculations for the porphine dianion (with a central hole radius of 1.97 Å) indicate that the ruffled structure is more stable than the planar structure. Work is now in progress to investigate the effect on the calculated quadrupole splitting of i) the ruffling of the porphinato core ; ii) a limited configuration interaction.

SUMMARY AND OUTLOOK

The size of the systems studied (about fifty atoms and more than two hundred electrons) has placed some severe constraints both on the models used and on the level of the calculations :

i) The systems considered are model systems which differ in several respects from the heme molecules found in the biological systems. They lack the substituents which are found at the periphery of the porphyrin ring in the heme. We have mimicked the imidazole ligand through an ammonia molecule (one calculation for an oxyheme model with an imidazole ligand indicates that this approximation may not be too severe[24] but we do not know how it may influence the out-of-plane displacement in the deoxyheme model). Idealized geometries were used with the porphyrin assumed to be planar (although, for instance, the four-coordinated FeTPP molecule is highly ruffled). We have ignored the possibility of doming of the porphinato ring by assuming a planar geometry (doming has not been established experimentally for the biological systems).

ii) The calculations for the oxyheme model have been carried out with a minimal basis set and, for instance, the calculated O-O bond length comes out too long (one calculation with a double-zeta basis set for the valence shells indicated that the qualitative outcomes from the minimal basis set calculations are probably correct [25]). Calculations for the deoxyheme model and the four-coordinated iron(II)porphyrin use a more flexible basis set for the valence shells but retain the minimal basis set for the core orbitals. Geometry optimization has been kept to a minimum. Correlation effects have been practically ignored although the correlation error is the source of difficulties in assigning the ground state of the oxyheme model. For this reason we have not attempted to calculate excitation energies (this would probably require much more flexible basis sets in order to have a balanced description of the ground and excited states [67]). A small number of configurations may be required in the ground-state wavefunction in order to achieve significant values of the electric field gradient and of the quadrupole splitting in the Mössbauer spectrum [19,21,26] (although they represent expectation values of one-electron operators).

Despite these limitations, we believe that the ab initio calculations for the model systems of heme complement the experimental studies with respect to :

i) the assignment of the ground state and of the corresponding electronic configuration ;

ii) the structure of the central unit made of the metal, the axial ligands and the pyrrole nitrogens, as defined through its geometry (bond lengths and bond angles), its conformation and the corresponding potential energy curves ;

iii) the interpretation of spectroscopic data such as the infrared frequencies and the Mössbauer quadrupole splitting and their relationship with the electronic configuration ;

iv) the influence of the protein on the heme.

Some of the limitations which we have mentioned are the consequences of technical limitations :

- even with a highly efficient program, the corresponding computation times remain relatively long (from a few hours for the minimal basis sets calculations to about twenty hours for the extended basis set calculations with an Univac 1110, a computer of moderate speed). Further progress in methodology, coding and/or computer technology may bring these numbers down and make for instance geometry optimization a relatively routine procedure for these molecules.

- limited storage abilities (in terms of high-speed storage) has put some restrictions on the open-shell SCF calculations (for the largest systems they were restricted to the half closed-shell type).

- the use of large basis sets of contracted functions gives rise to extremely long lists of integrals (up to sixty millions). Most of the time of the SCF iterations is devoted to setting up the Fock matrix from this list of integrals.

Clearly such theoretical studies are undoubtedly linked to both the computer technology and the state of the art of SCF and CI programs. Attempts to go beyond the SCF level have been made at the semi-empirical level (Parr-Pariser-Pople and CNDO approximations [19,21]) for analogous models and at the ab initio level for much simpler models [23]. The present bottleneck in the two-electron integrals transformation problem will probably prevent for some time large scale CI calculations for realistic models. A limited CI or better a limited MC-SCF calculation are probably within reach. The use of pseudopotentials to describe the inner-shells will help in this respect (although the benefit expected is limited since all the atoms except one are light atoms).

ACKNOWLEDGMENTS

The authors wish to express their gratitude to Drs. M. Bénard, J. Demuynck, A. Strich and H. Veillard who have contributed to the developments in computational methods, making possible the studies reported above. They also want to thank the staff of the Centre de Calcul du C.N.R.S. in Strasbourg-Cronenbourg. Support from the C.N.R.S. through the A.T.P. no. 2240 is gratefully acknowledged.

REFERENCES

1. For a general review about the structure and mechanism of hemoglobin, see for instance : Perutz, M.F. : 1976, Br. Med. Bull., 32, p. 195.
2. Pauling, L. : 1949, "Hemoglobin", Butterworth, London, p. 57.
3. Griffith, J.S. : 1956, Proc. Roy. Soc. Ser. A, 235, p. 23.
4. Pauling, L. and Coryell, C.D. : 1936, Proc. Natl. Acad. Sci. USA, 22, p. 210.
5. Cerdonio, M., Congiu-Castellano, A., Mogno, F., Pispisa, B., Romani, G.L. and Vitale, S. : 1977, Proc. Natl. Acad. Sci. USA, 74, p. 398.
6. Fermi , G. : 1975, J. Mol. Biol., 97, p. 237.
7. Shulman, R.G., Eisenberger, P., Kincaid, B.M., Brown, G.S. and Teo, B.K. : 1978, Abstracts, 22nd Meeting of the Biophysical Society, Washington, D.C., March 27-30 ; 1978, Biophys. J., 21, p. 173a.
8. Collman, J.P. : 1977, Acc. Chem. Res., 10, p. 265.
9. Jameson, G.B., Molinaro, F.S., Ibers, J.A., Collman, J.P., Brauman, J.I., Rose, E. and Suslick, K.S. : 1978, J. Am. Chem. Soc., 100, p. 6769.
10. Jameson, G.B., Rodley, G.A., Robinson, W.T., Gagne, R.R., Reed, C.A. and Collman, J.P. : 1978, Inorg. Chem., 17, p. 850.
11. Celotta, R., Bennett, R., Hall, J., Siegel, M. and Levine, J. : 1972, Phys. Rev., A6, p. 631.
12. Collman, J.P., Brauman, J.I., Halbert, T.R. and Suslick, K.S. : 1976, Proc. Natl. Acad. Sci. USA, 73, 3333.
13. Dedieu, A., Rohmer, M.-M. and Veillard, A. : 1976, J. Am. Chem. Soc., 98, p. 5789.
14. Collman, J.P., Gagne, R.R., Reed, C.A., Halbert, T.R., Lang, G. and Robinson, W.T. : 1975, J. Am. Chem. Soc., 97, p. 1427.
15. Collman, J.P., private communication. See also Ref. 39 of Ref. 8.
16. Lang, G. : 1970, Quart. Rev. Biophys., 3, p. 1.
17. For a review, see : Gouterman, M., 1978, in "The porphyrins", D. Dolphin ed., Academic Press, New York, 1978, Vol. 3.
18. Kirchner, R.F. and Loew, G.H. : 1977, J. Am. Chem. Soc., 99, p. 4639.

19. Huynh, B.H., Case, D.A. and Karplus, M. : 1977, J. Am. Chem. Soc., 99, p. 6103.
20. Loew, G.H. and Kirchner, R.F. : 1978, Biophys. J., 22, p. 179.
21. Herman, Z.S. and Loew, G.H., private communication.
22. Goddard, W.A. and Olafson, B.D. : 1975, Proc. Natl. Acad. Sci. USA, 72, p. 2335.
23. Olafson, B.D. and Goddard, W.A. : 1977, Proc. Natl. Acad. Sci. USA, 74, p. 1315.
24. Dedieu, A., Rohmer, M.-M. and Veillard, A. : 1977, in "Metal-ligand interactions in organic chemistry and biochemistry", B. Pullman and N. Goldblum eds., D. Reidel, Dordrecht, Netherlands, part 2, p. 101.
25. Dedieu, A., Rohmer, M.-M., Veillard, H. and Veillard, A. : 1979, Nouv. J. Chim., in the press.
26. Dedieu, A., Rohmer, M.-M. and Veillard, A., to be published.
27. Bénard, M., Dedieu, A., Demuynck, J., Rohmer, M.-M., Strich, A. and Veillard, A., "Asterix, a system of programs for the Univac 1110", unpublished work.
28. Hehre, W.J., Steward, R.F. and Pople, J.A. : 1969, J. Chem. Phys., 51, p. 2657.
29. Bénard, M. : 1976, J. Chim. Phys., p. 413.
30. Bénard, M. and Barry, M., Comp. in Chem., in the press.
31. Veillard, A. : 1975, in "Computational Techniques in Quantum Chemistry and Molecular Physics", G. Diercksen, B. Sutcliffe and A. Veillard eds., Reidel, Dordrecht, Netherlands, p. 201.
32. Koster, A.S. : 1975, J. Chem. Phys., 63, p. 3284.
33. Phillips, S.E.V. : 1978, Nature, 273, p. 247.
34. Weber, E., Steigemann, W., Jones, T.A. and Huber, R. : 1978, J. Mol. Biol., 120, p. 327.
35. Hoard, J.L. : 1978, in "Structural chemistry and molecular biology" A. Rich and N. Davidson eds., W.H. Freeman, San Francisco, 1968, p. 573.
36. Guilard, R., Fontesse, M., Fournari, P., Lecomte, Cl. and Protas, J. : 1976, J.C.S. Chem. Comm., p. 161.
37. Chevrier, B., Diebold, Th. and Weiss, R. : 1976, Inorg. Chim. Acta, 19, p. L57.
38. Bachmann, C., Demuynck, J. and Veillard, A. : 1978, J. Am. Chem. Soc., 100, p. 2366.
39. Schaefer, H.F. : 1974, "Critical evaluation of chemical and physical structure information", D.R. Lide and M.A. Paul eds., National Academy of Sciences, Washington, 1974.
40. Hoard, J.L. : 1975, in "Porphyrins and metalloporphyrins", K.M. Smith, ed., Elsevier, Amsterdam, 1975, p. 317.
41. Dedieu, A., Rohmer, M.-M., Veillard, H. and Veillard, A. : 1976, Bull. Soc. Chim. Belg., 85, p. 953.
42. Newton, M.D., Lathan, W.A., Hehre, W.J. and Pople, J.A. : 1970, J. Chem. Phys., 52, p. 4064.
43. Rohmer, M.-M., Barry, M., Dedieu, A. and Veillard, A. : 1977, Int. J. Quant. Chem., Quantum Biology Symposium, 4, p. 337.
44. Pauling, L. : 1964, Nature, 203, p. 182.

45. Antonini, E. and Brunori, M. : 1971, "Hemoglobin and myoglobin in their reactions with ligands", North-Holland Publ., Amsterdam, p. 85 and following.
46. Collman, J.P., Brauman, J.I., Suslick, K.S. : 1975, J. Am. Chem. Soc., 97, p. 7185.
47. Huynh, B.H., Papaefthymiou, G.C., Yen, C.S., Groves, J.L. and Wu, C.S. : 1974, J. Chem. Phys., 61, p. 3750.
48. Kent, T., Spartalian, K., Lang, G. and Yonetani, T. : 1977, Biochim. Biophys. Acta, 490, p. 331.
49. Jameson, G.B., Robinson, W.T., Collman, J.P. and Sorrell, T.N. : 1978, Inorg. Chem., 17, p. 858.
50. Perutz, M.F., Heidner, E.J., Ladner, J.E., Beetlestone, J.G., Ho, C. and Slade, E.F. : 1974, Biochemistry, 13, p. 2187.
51. Takano, T. : 1977, J. Mol. Biol., 110, p. 569.
52. Gelin, B.R. and Karplus, M. : 1977, Proc. Natl. Acad. Sci. USA, 74, p. 801.
53. Eisenberger, P., Shulman, R.G., Brown, G.S. and Ogawa, S. : 1976, Proc. Natl. Acad. Sci. USA, 73, p. 491.
54. Spiro, T.G. and Burke, J.M. : 1976, J. Am. Chem. Soc., 98, p. 5482.
55. Perutz, M.F. : 1978, Scientif. Amer., 239 (6), p. 68.
56. Cotton, F.A. and Wilkinson, G. : 1972, "Advanced Inorganic Chemistry", Interscience Publ., New-York, p. 871.
57. Hoard, J.L. : 1975, in "Porphyrins and metalloporphyrins", K.M. Smith ed., Elsevier, Amsterdam, p. 352.
58. Collman, J.P., Hoard, J.L., Kim, N., Lang, G. and Reed, C.A. : 1975, J. Am. Chem. Soc., 97, p. 2676.
59. Goff, H., La Mar, G.N. and Reed, C.A. : 1977, J. Am. Chem. Soc., 99, p. 3641.
60. Dale, B.W., Williams, R.J.P., Edwards, P.R. and Johnson, C.E. : 1968, J. Chem. Phys., 49, p. 3445.
61. Lang, G., Spartalian, K., Reed, C.A. and Collman, J.P. : 1978, J. Chem. Phys., 69, p. 5424.
62. Zerner, M., Gouterman, M. and Kobayashi, H. : 1966, Theoret. Chim. Acta, 6, p. 363.
63. Kitagawa, T. and Teraoka, J. : 1979, Chem. Phys. Let., 63, p. 443.
64. Moller, C. and Plesset, M.S. : 1934, Phys. Rev., 46, p. 618.
65. Huo, W. : 1965, J. Chem. Phys., 43, p. 624.
66. Mcweeny, R. and Sutcliffe, B.T. : 1969, "Methods of molecular quantum mechanics", Academic Press, London, p. 47.
67. Demuynck, J., Veillard, A. and Wahlgren, U. : 1973, J. Am. Chem. Soc., 95, p. 5563.

CALCULATIONS ON THE MECHANISM OF ION POLYMERIZATION IN CRYSTALS

M.V.Basilevsky, G.N.Gerasimov, S.I.Petrochenko, V.A.Tikhomirov
Karpov Institute of Physical Chemistry,
107120, Moscow B-120, USSR

The results of the quantum-chemical investigation of the radiation-induced polymerization in crystals are presented. The main elementary reaction steps are detected, giving a reasonable interpretation of the kinetic data. The basic conclusion is that the propagation reaction involving the addition of the growing polymer ion to the monomer molecule placed at a lattice point needs no activation energy. The experimentally observed activation energy is attributed to conformational processes involved in the termination stage.

1. INTRODUCTION

Experimental investigations of the radiation-induced solid phase polymerization have revealed a remarkable phenomenon. Its essence is illustrated in Fig.1. The regular temperature dependence of the polymerization rate was

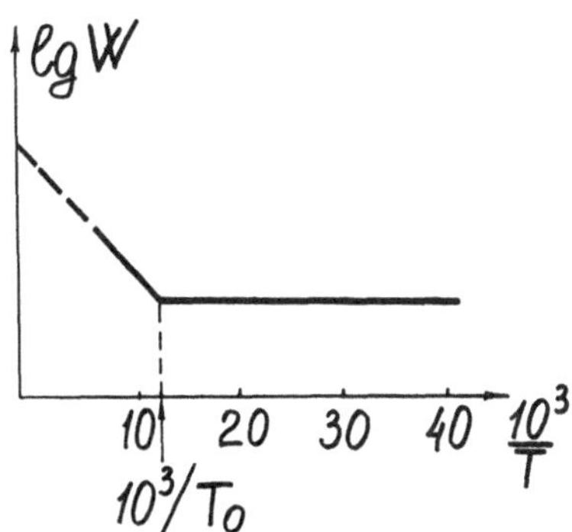

Figure 1. Arrhenius plot for the polymerization of formaldehyde[1] (w is the polymerization rate). $T_o \simeq 80°K$; activation energy $E_{eff}=2\pm0.5$ kcal/mole.

K. Fukui and B. Pullman (eds.), Horizons of Quantum Chemistry, 227–243.

observed at the temperature above 80-100°K, the activation energy being 2-3 kcal/mole. The regularity however changed drastically at low temperature: the process became activationless and occured even at the liquid helium temperature of 4.2°K. An appreciable break of the rate temperature dependence has been reported for polymerization of crystals of several vinyl monomers (tetrafluoroethylene acrylonitrile [2,3,4] , acrolein [5]) and also for formaldehyde [1].

The chain propagation step was shown to involve invariably ionic centres [6,7] . The unusual temperature dependence of the polymerization rate was a subject of the extensive discussion [8-10] . It was just the propagation step that was always accepted to cause the activation behaviour of the high-temperature kinetics. That is to say, the observed activation energy was associated with the addition reaction of a growing polymer ion R^+ and monomer molecule M:

$$R^+ + M \longrightarrow R^+$$

Thereby the basic observation to be accounted for was the low-temperature branch of the Arrhenius plot depicted in Fig. 1. It was accepted that apart from the regular mechanism, following the Arrhenius kinetics, an alternative low-temperature mechanism existed, that needed no Arrhenius activation for passing the potential barrier at the propagation step. The critical discussion of various versions of the latter mechanism presented by different authors may be found in several reviews [7,10] . The most recent idea suggested the quantum-mechanical tunnelling of large monomer molecules M to be responsible for the low-temperature transitions [10] . Our opinion is that this hypothesis is also an unsatisfactory one, since the tunneling probabilities, as estimated for real polymerizable systems, are extremely small and cannot account for the observed reaction rates.

The theme of the present communication is a new mechanism that we propose to ionic polymerization in molecular crystals. It is based on quantum-chemical calculations on a model system. Contrary to the preceeding authors we assume that the propagation reaction in a crystal has no potential barrier and thereby needs no activation energy, just as in the gas phase. That is a consequence of the high level of organization, specific for the monomer crystals in which polymerization occurs. Our main assumption makes self-evident their low-temperature activationless behaviour; however, a reasonable explanation is then needed for the slope of the high-temperature branch of the Arrhenius plot. This explanation is as follows. Several elementary chemical stages contribute to the observed overall reaction rate: not only the propagation reaction, but also the processes of generation and termination of polymer chains. The termination reaction is associated with the spontaneous conformational changes of the growing polymer ion turning it into an inactive form. An increase in temperature stimulates the reactivation from such conformational traps. Thus the observed activation energy may be associated with the reverse conformational transitions of the polymer chain.

2. THE CALCULATION METHOD

We investigated potential energy surfaces or, to be more precise, their several regions, for the ionic addition reactions proceeding under the influence of the potential that is generated by a crystal built of monomer molecules. We also calculated several conformations of the polymer ion fragments both in the gas phase and in a crystal. Extensive geometry optimization was found to be necessary, which made the calculations rather complicated. The calculations of the overall interaction energy were performed for an idealized model system, the semiempirical MO treatment being combined with the empirical atom-atom potential scheme.

The ethylene crystal chosen as a model system is schematically sketched in Fig.2. It forms a rectangular

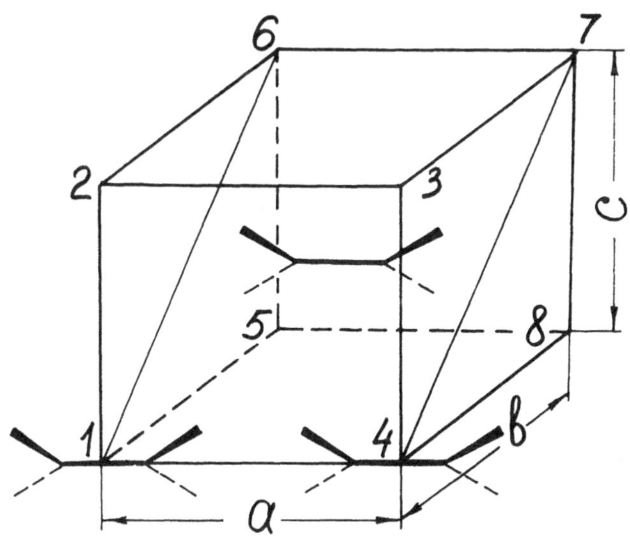

Figure 2. Unit cell of the model ethylene crystal. a = 5.04 Å , b = 3,93Å, c = 6.71 Å

body-centered lattice. The reaction proceeds in the plane 1674 of the unit cell. The ethylene molecules are oriented in this plane in the same direction coinciding with the direction of the 14 edge. The molecular planes are perpendicular to the reaction plane. Optimization of the lattice distances a, b, c was performed by a regular procedure [11]. The orientation of molecules at lattice points as well as the unit cell angles were not optimized, being adjusted beforehand. There is a conclusive experimental evidence that polymerization occurs only in those crystals, where the orientation and disposition of monomer molecules are most favourable for the successive addition reactions [7]. The set of reacting molecules in the crystal forms a sort of pattern, the prototype of the future polymer chain. Note that the molecular packing in the real ethylene crystal distinguishes from our model, but it is known that the ethylene cannot polymerize in the crystalline form. A special choice of our model crystal is such, that it involves patterns accelerating the reaction and resembling those existing in real polymerizable crystals.

Fig.3 clarifies the process of polymerization in a pattern constructed of the ethylene molecules fixed in the reaction plane along a pair of parallel axes AA′ and BB′.

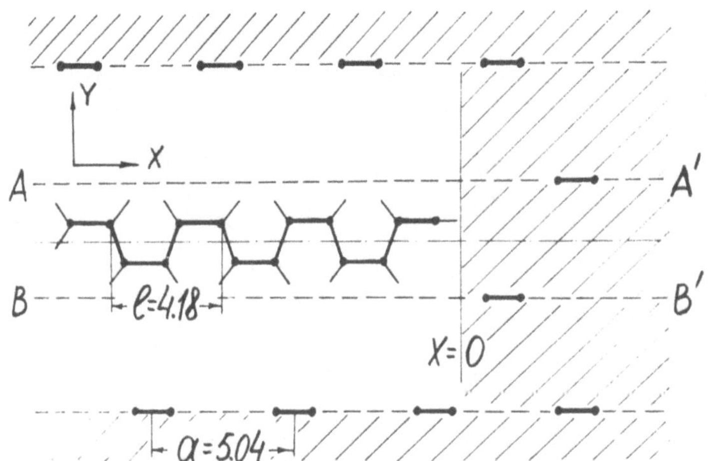

Figure 3. Scheme of polymerization.
The molecules of the pattern form the polymer chain, the "corridor" (the unshaded region) arising in their place. a is the lattice period, ℓ is the period of the chain (in Å).

We considered only two external degrees of freedom of the reacting molecules, namely, their parallel translations in the reaction plane. The reaction produces the polymer chain with a flat carbon backbone in cis-conformation. The molecules surrounding the pattern were fixed in their positions at lattice points. The same was accepted for the pattern molecules other than a reacting pair. We shall refer to this model as to a "rigid chain - rigid lattice" approximation.

The main outcome of the calculations were potential curves characterizing different molecular motions in the crystal, including chemical reactions. In the latter case their calculation proceeded as follows. We selected a reacting subsystem, whose internal interaction energy, the chemical energy U_{chem} , was calculated using a conventional quantum-chemical procedure. Another contribution added to this energy was the crystalline energy U_{cryst} , representing the interaction of the reacting subsystem with the motionless crystalline background. The latter term was calculated by the empirical atom-atom potential scheme. The calculation of the overall interaction energy,

$$U = U_{chem} + U_{cryst} \tag{1}$$

involved optimization of both the internal degrees of freedom of the subsystem and its position, on the reaction plane inside the crystal under the constrints mentioned above. The quantum-chemical methods that we used were MINDO-3 and CNDO with a specific parametrization as suggested by Boyd and Whitehead [12] and refined by Lebedev and Bagaturyantz [13] (CNDO-BWL). The atom-atom calculation used parametrization due to Williams [14] .

The following motivation underlied our choice of the quantum-chemical method. The simplest of propagation reactions that we considered was

$$C_2H_5^+ + C_2H_4 \longrightarrow C_4H_9^+ + 30 \text{ kcal/mole}$$

The listed experimental value for the heat effect corresponds both to a flat classical configuration of the product, the butyl cation $C_4H_9^+$, and to a classical configuration of the reactant ethyl cation (its estimation is based on the experimental vertical ionization potentials the respective radicals having classical structures).That are just those conformations involved in our model reaction (Fig.3). The two conditions which we considered to be obligatory for a satisfactory calculation of the above reaction in the gas phase, i.e. in absence of the crystal field, were, first, that it must yield a satisfactory heat effect, and, second, that the calculated potential curve must be a smooth one, without a potential barrier. The results of different calculations are presented in Fig.4.

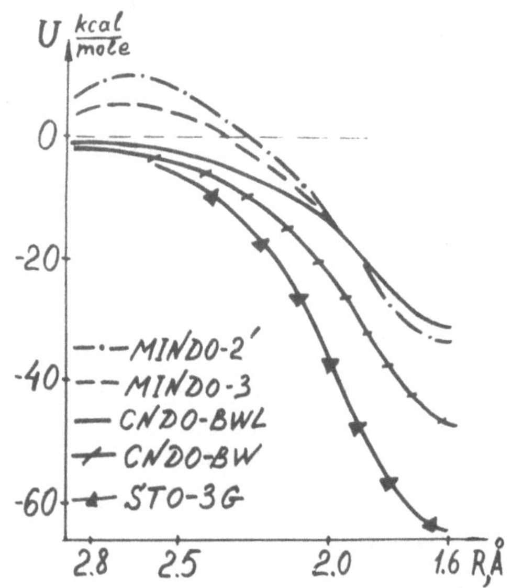

Figure 4. The potential curves for the $C_2H_5^+ + C_2H_4$ addition reaction. The experimental heat effect equals to 30 kcal/mole. R is the length of the new C-C bond. MINDO-3 and CNDO-BWL curves coincide for small R values.

The STO-3G calculation gave no barrier but it greatly exaggerated the heat effect. Both MINDO-3 and MINDO-2 gave an appropriate heat effect but the appearance of a potential barrier made them useless. The regular CNDO-BW procedure resulted in no barrier but the heat effect became unsatisfactory again. Finally, the CNDO-BWL was found to be most fit for our purpose, giving a smooth potential curve with a satisfactory heat effect of 31 kcal/mole. So we used CNDO-BWL in the calculation of addition reactions. For conformational calculations we used MINDO-3, the method, well reproducing both the geometries and relative energies of different conformers of hydrocarbon cations [15].

3. GENERATION OF THE POLYMER CHAINS

The polymerization starts with the emergence of a

primary ion M^+ at a lattice point. As such an ion we considered the cation $C_2H_5^+$, the product of proton addition to ethylene. This is a possible way of the chain initiation under radiation.

Then the primary generation reaction is

$$C_2H_5^+ + C_2H_4 \longrightarrow C_4H_9^+$$

and it is illustrated in Fig.5. The origin of the potential barrier displayed in Fig.6a becomes quite clear by

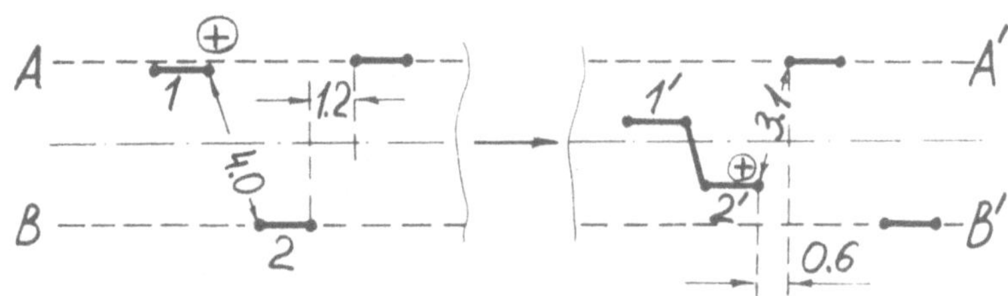

Figure 5. Scheme of the generation reaction. The numbers 1, 2 and 1′, 2′ label the positions of reactants before and after the reaction respectively. All distances in Å .

analysing how the two terms in formula (1) behave in the course of the reaction. At the initial stage an energy expense is necessary for overcoming the crystalline forces that retain the reactants at lattice points. On the other hand chemical energy U_{chem} decreases smoothly along the reaction path (the curve CNDO-BWL of Fig.4). As a result of rapid damping of U_{chem} with the increase of R, the length of the new C_2-C_3 bond, the chemical energy release becomes noticeable only at the later stage, when the bond becomes sufficiently strong. As a matter of fact, it is the competition of U_{cryst} and U_{chem} which produces a potential barrier within the intermediate range of R.

It is not clear yet how the system gets across this primary barrier at low temperature. As the most probable mechanism we consider an activationless transition promoted by a vibrational excitation of the primary ion M^+ at the

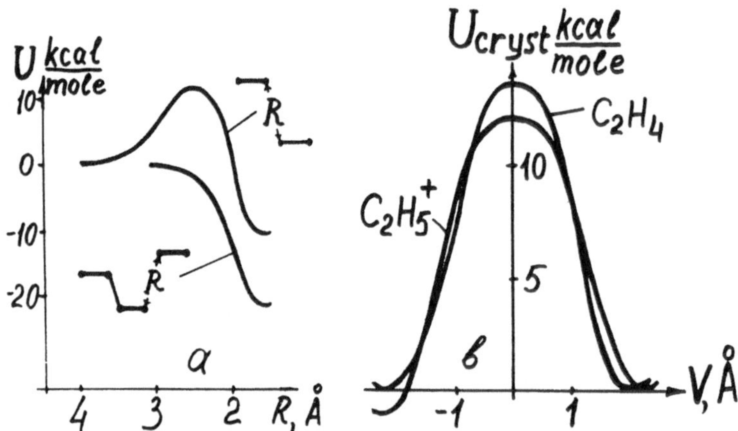

Figure 6. Potential curves for the generation and propagation reactions. (a) The overall energy profiles along the reaction coordinate. (b) Crystalline energy of reactant molecules.

moment of its formation at a lattice point [16]. Still, this suggestion is deficient in at least one respect. The transfer of the vibrational energy into the reaction channel occurs on the background of the extremely fast process of its relaxation into the crystal. That is why the efficiency of such a way of the chain generation is expected to be quite small, much less than that observed in the experiment. So some unknown mechanism should exist, localizing the vibrational excitation for a proper time in the immediate neighbourhood of the cationic centre.

The crystalline energy is minimal for the course of the primary addition presented in Fig.5. Most important is the result that the product $C_4H_9^+$ inserts into the space between the lattice layers. The following reasoning is chiefly based on this conclusion, so we have deduced it from a general consideration without any reference to calculational details [16] . The proof made use of general properties of the crystalline potential U_{cryst} . For the combined reacting system $C_4H_9^+$ it was a sum of the potentials contributed by its fragments $C_2H_5^+$ and C_2H_4 . The potential curves in Fig.6b correspond to the motions of each of them from the point of its location before the reaction to that, where its partner has been located. The interaction with the partner was extracted out of U_{cryst}, being incorporated into a quantum-chemical calculation of U_{chem}. Hence the curves in Fig.6b represent a motion inside a crystal having the defect, the vacancy in the place of the partner. They are quite similar owing to a close geometrical resemblance of the reacting fragments. That is why the overall crystalline potential has certain symmetry properties resulting in the fact of stabilization of the combined $C_4H_9^+$ system just in the middle of the pattern between the adjacent lattice points where the reactants have been located.

4. PROPAGATION OF THE POLYMER CHAINS

Thus we have found an optimal position of the cation $C_4H_9^+$ in the crystal provided the chemical energy of its interaction with the next monomer molecule is switched off. The distance R between the cationic centre and the closest to it carbon atom of the ethylene at an adjacent lattice point amounted to 3.1 Å . Given this initial position, the chemical interaction was sitched on and the next addition reaction,

$$C_4H_9^+ + C_2H_4 \longrightarrow C_6H_{13}^+$$

was calculated. No barrier was found along the reaction path (fig.6a).

Now it suffices to demonstrate that the position of the crystalline energy minimum respective to the remainder of the pattern is almost the same for the whole family of polymer ions R^+ of arbitrary length. If so, no activation energy would be needed at any propagation step.

During the propagation process the two monomer layers of the pattern convert into the growing polymer ion, the "corridor" arising in their place. We investigated motions of different molecular fragments along that semi-infinite corridor shown in Fig.3. The coordinate X represented the position along the central axis of the terminal carbon atom of a moving fragment. The transversal motions were optimized, with the result that the system was always retained close to that axis.

Again only crystalline energy was varied in the calculation. Due to the atom-atom scheme it was additive, and the potential of the whole chain was a sum of the fragment potentials. Fig.7 reproduces the potential curves for the cationic fragment $C_4H_8^+$, the head of the polymer ion, and for the repeating uncharged fragment C_4H_8 of the rest of the chain. The origin x=0 was assigned to the energy minimum of $C_4H_8^+$. Far off that border the motion proceeded under the periodic potential. Strictly speaking, its period λ should coincide with the lattice period a . However, for the linear systems (more precisely, for the systems, whose projections on the reaction plane are linear) arranged along the central axis the period becomes twice as small, $\lambda = a/2$. Our fragments were not exactly linear ones, but their width (~2Å) was much smaller than the height of the corridor (~12Å). The consequence was that their potentials could be approximately considered as harmonic ones with the period $\lambda = a/2$, the corresponding error being negligibly small. Note that the potential amplitude for C_4H_8 is approximately twice as small as that for $C_4H_8^+$.

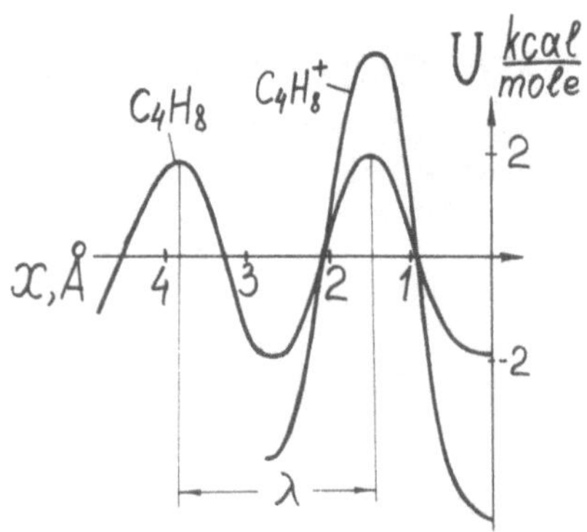

Figure 7. Potential curves for the motion of the fragments in the "corridor" (see Fig.3).

What remains to do is summation of the fragment potentials. First we shall obtain the overall potential of the uncharged "tail" of the polymer ion. Its value depends drastically on the ratio of the periods of the fragment potential (λ) and the chain (ℓ), which determines a degree of mutual compensation of several fragment contributions. For harmonic fragment potentials the summation is performed analytically, giving the overall potential for the chain built of N dimeric fragments:

$$U_N(x) = A_N \sin[2\pi(x+x_N)/\lambda]\ ;\ A_N = A \sin(N\pi\ell/\lambda)/\sin(\pi\ell/\lambda) \quad (2)$$

Here A represents the amplitude of the fragment potential, A_N and x_N being the N-dependent amplitude and phase shift of the overall sum. It follows from formula (2) that the value of the overall amplitude A_N is of orderof A provided $\ell/\lambda \approx (2n+1)/2$ (n = 0,1,2,...), whereas it becomes of order of NA, i.e. is proportional to the chain length, if $\ell/\lambda \approx n$. In our model $\ell/\lambda = 1.66$, so the compensation was almost complete. The amplitude A_N was found to be less than the amplitude of the head fragment $C_4H_8^+$ for N varying from 1 up to 10^3.

We conclude that the character of the motion does not change after attaching the "tail" to the "head". For N varied between 1 and 10^3 the optimal position of the polymer ion R^+ near the border varied within 0.3 Å , being such that the following addition reaction needed no activation.

In our model the polymer ion should be translated by the interval of d≈0.5 Å at every addition step. Then the initial position of the preceeding step will be reproduced. Since $d \ll \lambda/2 = 1.25$ Å , so after every addition act the ion is found in the slope region and it moves spontaneously to the desired energy minimum. The chain makes its next in turn step, takes the optimal position for the next in turn activationless addition act and so on.

This effect is the most remarkable result of our calculation. The chain of the length 2N = $2\cdot10^3$ was spontaneously transferred over the distance of 1000 Å!

5. TERMINATION OF THE POLYMER CHAINS

Basing on the above results we can now develope the following kinetic scheme.

(O)	$M \xrightarrow{h\nu} M^+$	v_o
(I)	$M^+ \xrightarrow{k_1} R^+$	$v_1 = k_1 m$
(I')	$M^+ \xrightarrow{k_1'} Z$	$v_1' = k_1' m$

$$(II)\quad R^+ + M \xrightarrow{k_2} R^+ \qquad v_2=0;\quad w=k_2 r$$

$$(III)\quad R^+ \xrightarrow{k_3} Z \qquad v_3=k_3 r$$

$$(IV)\quad R^+ \underset{k_{-4}}{\overset{k_4}{\rightleftharpoons}} Y^+ \qquad v_4=k_4 r;\quad v_{-4}=k_{-4} y$$

$$(IV')\quad Y^+ \xrightarrow{k_4'} Z \qquad v_4'=k_4' y$$

The notation: m,r,y are respectively the concentrations of the primary monomer ions M^+ , the active polymer ions R^+ and the intermediate inactive ions Y^+. The inactive products of irreversible reactions are denoted by Z. The rates v_i correspond to concentration changes of various ions, w is the observed rate of formation of the polymer chain links or, which is the same, the rate of vanishing of monomer molecules. Reaction 0 represents an initiation step, its main result being the formation of primary monomer ions in lattice points. Reaction I represents the desactivation of primary ions. Reactions III, IV and IV′ are different termination processes. The rate expression for stationary regime is

$$w = v_1\, k_2/K_3 \tag{3}$$

with the effective termination rate constant:

$$K_3 = k_3 + k_4 k_4'/(k_{-4}+k_4') \tag{4}$$

Let us specify in more detail the termination reactions. Reaction III is an irreversible termination due to the recombination of ions and electrons. This process was shown to be extremely slow ($k_3 \approx 10^{-1} - 10^{-2}\ s^{-1}$ [17]), so it seems to be negligible. Reaction IV at high temperature corresponds to conformational equilibrium, IV′ is an irreversible degradation of the intermediate conformers.

Conformational transitions occur due to instability of hydrocarbon cations of classical structure. Being isolated, they undergo a spontaneous rearrangement into various nonclassical structures corresponding to local energy minima [15] . The classical cis-conformation of the polymer ion becomes metastable inside a crystal, because it is retained due to the interaction with the crystalline environment. As shown in Sec.4, the chain propagation (rate constant k_2) is an activationless process. The rate of the chain generation v_1 also shows only a weak temperature dependence [16] . Therefore formula (3) tells us that the observed activation energy E_{eff} of the polymerization rate w is mainly due to the termination reaction with the activation energy E_3:

$$E_{eff} \approx -E_3$$

Since the observed E_{eff} amounts to 2-3 kcal/mole, so we have to anticipate the possibility of small negative values for E_3.

Let us consider for that purpose the conformational contribution into (4)

$$K_{conf} = k_4 k_4' / (k_{-4} + k_4') \qquad (5)$$

Note that purely equilibrium transitions ($k_4' = 0$) do not influence the overall kinetics. We obtain a negative temperature dependence for K_{conf} by assuming k_4 and k_4' to be temperature independent. Thereby

$$K_{conf} = \begin{cases} k_4 & \text{(low temperature)} \\ k_4 k_4'/k_{-4} & \text{(high temperature)} \end{cases} \qquad (6)$$

At high temperature $k_{-4} \gg k_4'$. Now we arrive at the conclusion that E_3 can be negative, and then according to (4) (with k_3 neglected) its absolute value will coincide with the heat of the conformational change IV ($E_3 \approx -E_4$) .

The conformational process shown in Fig.8 is a model illustration. Our MINDO-3 calculation of butyl cation have revealed a local energy minimum for the nonclassical structure B of the cyclopropyl cation type with $C_1C_2C_3$ angle ~70°

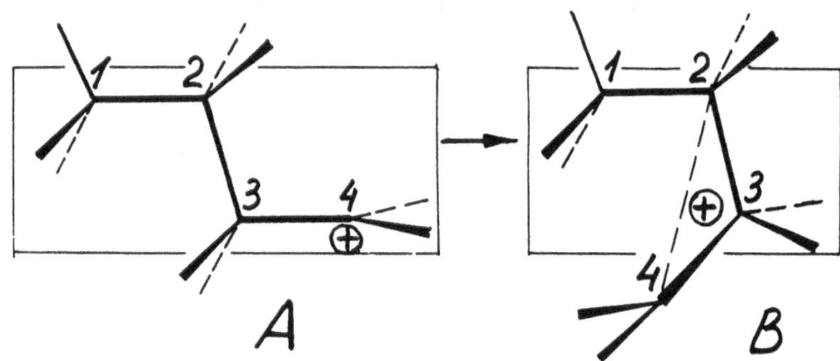

Figure 8. Conformational change of the butyl cation. The bond lengths for the nonclassical structure B are R_{23}=1,54 Å , R_{34} = 1,45 Å , R_{24} = 1.72 Å.

and C_1-C_2 bond turned by 60° out of the reaction plane around the fixed bond C_2-C_3. This is also consistent with a STO-3G calculation [18] . Similar nonclassical conformations have been reported for the propyl cation [15,19]. So in the gas phase the classical structure A is unstable and converts into B without activation. It is however stabilized inside a crystal, because the side walls of the "corridor" hinder the internal rotation around the bond C_2-C_3. The A → B transition can be performed only until the head of the polymer ion, while moving along the corridor, meets a specific crystalline defect, the vacancy in a side wall. Then a spontaneous conformational transition occurs, and the chain comes to a stop. At low temperature that means an irreversible termination. At high temperature a reactivation is possible, which results in an effective activation energy appearing according to formulas (4)-(6).

6. DISCUSSION

Our calculations are imperfect by several reasons and their validity may need verification. From that point of view their results may be divided into two categories.

(a) We consider as reliable the conclusion concerning the activationless character of the propagation reaction. Here we were able to detect a physical mechanism underlying the numerical results. It follows from general properties of the crystalline field and depends neither on the simplifications of the model nor on the calculation parameters.

(b) There is no doubt that the barrier calculations for both the generation and termination reactions should be refined. For instance, the barrier found for the primary reaction I ($E_1 > 10$ kcal/mole) is certainly too high. We consider it as a deficiency of the model "rigid chain-rigid lattice" The energy expenses at the initial stage could be significantly lowered if we optimized additional degrees of freedom, such as rotations of reactant molecules in the reaction plane. On the other hand the barriers of termination reactions are very sensitive to the quality of quantum-chemical calculations. So the refinement of the treatment of E_{chem} could significantly change the results concerning conformational processes.

The kinetic scheme proposed in Sec.5 seems to be useful as a guideline. The low-temperature mechanism, that it suggests, is universal, being based on general principles common to all polymerizing crystals. It is only the nature of the activationless transition at the primary chain generation stage, that is unclear as yet. On the other hand, the conclusions concerning the nature of low-temperature processes are much less definite. The termination reactions are very specific, because the conformational changes are expected to be different in different systems. In order to ascertain them the new complicated qualitative calculations of nonclassical conformations in crystals are quite desirable.

7. CONCLUSION

The most interesting results of the present work proved to be a consequence of the high degree of organization of the system under study. First, the polymerizing crystals involve patterns providing spacial conditions for the sequential propagation reactions that are close to the optimum ones. (The idea of "patterns" was suggested by Semyonov in 1960 [20]). Second, the periodic character of the crystalline potential together with the periodicity of the polymerization process make an ideal combination. By this a chemical machine is created, which is able of transferring a polymer chain, the object comparable in complexity with biological systems, over macroscopic distances of thousands Angströms.

The biological analogies are in order. In particular, the stated above mechanism of the polymer chain translation may be of some interest in connection with the problem of muscle contraction.

From the professional point of view of a quantum chemist the main specific feature of our treatment was partitioning of the interaction energy into the "chemical" and "crystalline" components according to formula (1). The former term is available from a standard quantum-chemical calculation for a rather small chemically active fragment. The latter one represents a "nonchemical" influence of the remote molecular fragments on a chemical reaction. Quantum-chemical procedures are not well fit to deal with the crystalline term by two reasons: (a) the methods usually available (semiempirical, in HF approximation) fail to reproduce correctly the long-range component of the interaction energy, and (b) the crystalline energy is generated by a combined action of a great number of remote atoms, this number being so large, that the diagonalization of the respective hamiltonian matrix seems to be an unrealizable task. Thus we conclude that the atom-atom approximation is inevitable and that the partitioning of the interaction energy into two complementary fractions is natural.

It is important to emphasize here that the chemical energy contribution proved to be trivial in our system, whose main distinctive features were just associated with the crystalline energy. This is not surprising. As pointed above, those features were a reflection of high organization of the system, which is essentially a macroscopic property, naturally associated with the macroscopic part of the potential operating a chemical reaction.

Here we again return to biological analogies. The main distinction of the enzyme catalysis is that a protein remainder significantly affects a reaction between a substrate and the active centre of the enzyme. But the protein can be considered as an "aperiodic crystal" [21] , so it seems instructive to extend onto enzyme reactions the approach found to be useful for the investigation of reactions in regular crystals, namely, the energy partitioning (1).

Therefore the calculations on the polymerization in organized systems seem to have much in common with theoretical problems of biopolymers, both in their ideology and methodology.

REFERENCES

1. Kiryukhin, D.P., Kaplan, A.M., Barkalov, I.M., and Goldansky, W.I.: 1972, Wisokomol. soedinenija, A14, p.2115 (in Russian).
2. Bruck, M.A., Gromov, W.F., Chernyak, I.W., Khomikovsky, P.M., and Abkin, A.D.: 1966, Wisokomol. soedinenija, 8, p.961 (in Russian).
3. Gerasimov, G.N., Bruck, M.A, Gromov, W.F., Abkin, A.D., Khomikovsky, P.M., Matveeva, A.V., and Chernyak, I.W.: 1967, Doklady Academii Nauk SSSR, 174, p.386 (in Russian)
4. Barkalov, I.M., Gareeva, D.A., Goldansky, W.I., Enikolopyan, N.S., and Berlin, Al.Al.: 1966, Wisokomol. soedinenija, 8, p.1140 (in Russian).
5. Finkelshtein, E.I., Abkin A.D., and Gorbatov E.Ja.: 1969, Chimija wisokikh energii, 3, p.446.
6. Dolotov, S.M., Gerasimov, G.N., and Abkin, A.D.,1968, International symposium for makromolecular Chemistry, Proceedings, Tashkent (in Russian).
7. Abkin, A.D., Sheinker, A.P., and Gerasimov, G.N.: 1973, Radiation chemistry of polymers, Moscow, "Nauka", (in Russian).
8. Finkelshtein, I.E., and Abkin, A.D., 1967, Doklady Academii Nauk SSSR, 174, p.887 (in Russian).
9. Abkin, A.D., and Gerasimov, G.N.: 1967, Wisokomol. soedineniya, A13, p.240 (in Russian).
10. Goldansky, W.I.: 1975, Uspekhi chimii, 44, p.2121 (in Russian).
11. Kitaigorodsky, A.I.: 1971, Molecular crustals, Moscow, "Nauka" (in Russian).
12. Boyd, R.J., and Whitehead, M.A.: 1972, J. Chem. Soc., Dalton Trans., p.73.
13. Lebedev, W.L., and Bagaturyantz, A.A.: 1978, Journal fisich. chimii, 52, p.1309 (in Russian).
14. Elliot, G.R., and Leroi, G.E.: 1973, J. Chem. Phys., 59, p.1217.
15. Lischka, H., and Köhler, H.-J.: 1978, J. Am. Chem. Soc., 100, p.5297.
16. Basilevsky, M.V., Gerasimov, G.N., Kitrossky, L.N., Petrochenko, S.I., and Tikhomirov, V.A.: Kinetika i catalis (in press, in Russian).
17. Lin, J., Tsuji, K., and Williams, F.: 1967, J. Chem. Phys., 46, p.4982.
18. We are grateful to Dr. J.Chandrasekhar (Erlangen) who made for us this STO-3G calculation.
19. Hehre, W.J., 1977, Mod. Theor. Chemistry, vol.IV, p.277, Plenum Press.
20. Semyonov, N.N., 1960, Chimija i technologiya polimerov, N 7-8, p.196.
21. Schrödinger, E., 1944, What is life, Cambridge.

AB-INITIO COMPUTATIONS OF METAL-PORPHINE COMPLEXES

Kimio Ohno
Department of Chemistry, Faculty of Science,
Hokkaido University, Sapporo, 060, Japan

Electronic structure of Co-, Cu-, Fe- as well as metal-free porphines is studied by LCAO-SCF and limited CI.

Four kinds of basis sets are used. They are minimal, minimal plus diffuse, partial double, and double zeta.

It is well to note that: (i) Koopmans' relation holds when an electron is removed from porphine, while the reorganization energy amounts to 6∿7 eV when an electron is removed from the metal d orbitals. (ii) The net charge of the metal atom is in the range 1.0∿2.1 irrespective of the metal and independent of the valency (or the oxidation number) and also of the spin state of the system.

The excitation energies and the oscillator strengths are first calculated by singly excited π-CI. For Co- and Cu-porphines, the calculated Soret band is too high by 2.3 eV. Inclusion of diffuse orbitals in the basis did not improve this result. More extensive CI with multiple reference configurations including singly and doubly π excitations was also carried out with little success. CI including σ-π excitations seems to be necessary.

A number of computer program packages, JAMOL3 (LCAO-SCF) and COMICAL2 (CI), which have been written in our laboratory, are herein described.

1. INTRODUCTION

The first ab-initio molecular orbital (MO) calculation on porphyrins was carried out by Almlöf on free base porphine [1]. Since then, a fair number of such calculations have been performed on porphyrins and metalloporphyrins. The main aim of these calculations is to elucidate the chemistry of the centrally coordinated metal and the porphyrin macrocycle in terms of their electronic structure. The chemistry, in turn, would be used to understand a variety of important biological

K. Fukui and B. Pullman (eds.), Horizons of Quantum Chemistry, 245–266.

functions performed by metalloporphyrins. These functions include the storage and transport of electrons and molecular oxygen, the decomposition of hydrogen peroxide and the activation of oxygen. It is true that semi-empirical calculations are more widely used for such purposes and are, at the present time, probably more useful. However, it is clear that semi-empirical calculations have their disadvantages. For example, choice of parameters is always a difficult problem and seems even arbitrary in some cases. The number of physico-chemical properties measured on metal porphyrins has increased tremendously recently. For example, the ionization energy of valence and inner electrons, the Mössbauer isomer shift and quadrupole splitting, the resonance Raman spectra have been added to traditional data on ultra violet and visible absorption spectra, fluorescence, phosphorescence, CD, MCD and ESR, NMR spectra with their hyperfine structures. It is the author's opinion that no semi-empirical computation can correlate all of these properties for a single system. In principle, a good ab-initio calculation can do this. The bottlenecks of ab-initio computations are being broken by rapid development of electronic computers themselves and by clever algorithms and efficient programming. However, it is still far from easy to carry out ab-initio computations on a large system such as metal porphines and the task is still at its beggining stage.

Among the presently available ab-initio MO calculations on metal porphines, the work of two groups is worth special mentioning since it is a series of calculations and constitutes an organized effort towards the aim mentioned above. Magnesium porphine and related systems are investigated by Christoffersen and others [2-5] by configuration interaction (CI) starting from a self consistent field (SCF) calculation with a basis of floating spherical Gaussian orbitals. Excitation energies to lower singlet and triplet states are their central interest. They obtain a linear relationship between experimental and computed excitation energies and they analyze the resulting CI wave functions in terms of individual electron jump(s). The other group is led by Veillard. The geometry and charge distributions of various metal-porphine complexes, especially of iron-porphine complexes are examined at the SCF level with a basis of contracted Gaussian-type orbitals (CGTO's) [6, 7]. More detailed and recent results will be reported by Veillard at this Congress.

In this paper, a review of the work carried out by our own group is presented. This paper is also a progress report since most of the calculations are in progress and not yet finished.

In Table 1, the molecules which have been calculated are collected with some pertinent information on the assumed symmetry, the level of calculation and the basis set. An abbreviation is used to indicate the quality of a basis set in Table 1 and the abbreviation is explained in Table 2. Calculated results of ionization potentials, excitation energies, and charge distribution in these molecules are presented and discussed in Sects. 2, 3, and 4, respecitively. Conclusions are given in Sect. 5.

Table 1. Outline of the systems studied and the calculation

Molecules [abbr.]	No. of Elect.	Sym.	Level of Calc.	Basis Set	No. of CGTO's	Ref.
Free base porphine [H_2P] cation, anion	162	D_{2h}	SCF +πCI(S,D)	Min.	134	[8]
Free base chlorine [H_2C] cation, anion	164	C_{2v}	SCF +πCI(S,D)	Min.	136	[8]
Co-porphine [CoP] cation (+1,+2), anion	187	D_{4h}	SCF +πCI(S)	Partial Dbl.(I)	184	[9]
Cu-porphine [CuP] cation, anion	189	D_{4h}	SCF +πCI(S)	Min. +diffuse	177	[10]
Fe-porphine [FeP] (A)	186	C_{4v}	SCF	Partial Dbl.(I)	184	[11]
(B)		D_{4h}	SCF	Dbl.	300	[12]
Fe-porphine pyridine [FeP(Py)]	228	C_{2v}	SCF	Partial Dbl.(I)	223	[13]
Fe-porphine pyridine CO[FeP(Py)(CO)]	242	C_{2v}	SCF	Partial Dbl.(II)	241	[14]

Table 2. Basis sets

Abbr.	Atom	CGTO's	Ref.
Min.	N,C	(7s3p)/[2s1p]	[15]
	H	(3s)/[1s]	[16]
Min. +diffuse	Cu	(9s5p3d)/[3s2p1d] +3s,1p,$1p_{\pi}$	[17]
	N,C	(7s3p)/[2s1p]+$1p_{\pi}$	[15]
	H	(3s)/[1s]	[16]
Partial	Co	(12s8p5d)/[6s5p3d]	[9]
Dbl.(I)	Fe	(12s8p4d)/[6s5p3d]	[17]
	N	(7s3p)/[3s2p]	[15]
	C	(7s3p)/[2s1p]	[15]
	H	(3s)/[1s]	[16]
Partial	Fe	(12s8p4d)/[8s6p2d]	[17]
Dbl.(II)	N,O,C(CO)	(7s3p)/[3s2p]	[15]
	C(others)	(7s3p)/[2s1p]	[15]
	H	(3s)/[1s]	[16]
Dbl.	Fe	(14s11p5d)/[8s6p2d]	[18]
	N,C	(8s4p)/[4s2p]	[19]
	H	(4s)/[2s]	[19]

In the appendix, some characteristic features of our SCF (JAMOL3) and CI (COMICAL2) program packages are explained.

2. IONIZATION POTENTIAL

Ionization potentials are calculated in two ways. One is to take simply the sign reversed orbital energy ε. This is Koopmans' relation in case of closed shells. In case of open shells, the ground for Koopmans' relation is more dubious but $-\varepsilon$ seems, at least in metal porphines, to give a rough estimate of the ionization potential if the reorganization energy is small. The second way is to carry out a separate SCF calculation for neutral and ionized systems and subtract the energy of the former from that of the latter. The result of this calculation is labeled as ΔSCF in Table 3 in which ε's for a few highest occupied and lowest unoccupied orbitals and ΔSCF values are collected and compared with observed values.

Table 3. Ionization potentials and electron affinities (in eV)

		Calculated		Observed
		Orbital energy	ΔSCF	
H_2P	$4b_{2g}(e_g)$	-0.58	—	—
$[^1A_g]$	$4b_{3g}(e_g)$	-1.05	1.36	—
	$5b_{1u}(a_{2u})$	-7.65	7.23	6.39 [20]
	$2a_u\ (a_{1u})$	-7.93	7.59	6.83 [20]
	$3b_{2g}(e_g)$	-10.12	9.57	7.42 [20]
H_2C	$7a_2\ (e_g)$	-0.04	—	—
$[^1A_1]$	$9b_1\ (e_g)$	-0.50	0.89	—
	$6a_2\ (a_{1u})$	-7.39	6.98	—
	$8b_1\ (a_{2u})$	-7.72	7.23	—
CoP	$5e_g$	0.10	0.10	—
$[^2A_{1g}]$	$1a_{1u}$	-6.82	$6.51(^3A_{1u})$	6.09 [20] CoOEP
	$5a_{2u}$	-7.12	$6.77(^3A_{2u})$	6.58 [20] CoOEP
	$2e_g\ [d_\pi]$	-14.37	$7.47(^3E_g)$	—
	$14a_{1g}[d_{z^2}]$	-15.31	$9.06(^1A_{1g})$	—
CuP	$5e_g$	-0.18	$0.38(^1E_g)$	—
$[^2B_{1g}]$	$1a_{1u}$	-7.11	—	6.66 [20] CuOEP
	$5a_{2u}$	-7.25	—	7.08 [20] CuOEP
	$8b_{1g}[d_{x^2-y^2}]$	-17.23	$9.88(^1A_{1g})$	—
	$1e_g\ [d_\pi]$	-17.03	$9.91(^3E_g)$	—

Table 3 [continued]

FeP(A)	$5e_g$	-4.49	—	—
$[^3A_{2g}]$	$1a_{1u}$	-6.90	$6.57(^4A_{2u})$	6.06 [20] FeOEPCl
	$5a_{2u}$	-7.16	$6.82(^4A_{1u})$	6.48 [20] FeOEPCl
	$15a_{1g}[d_{z^2}]$	-11.09	$5.56(^4A_{2g})$	—
FeP(B)	$5e_g$	-4.30	—	—
$[^3A_{2g}]$	$1a_{1u}$	-6.40	$5.99(^4A_{2u})$	6.06 [20] FeOEPCl
	$5a_{2u}$	-6.90	$6.46(^4A_{1u})$	6.48 [20] FeOEPCl
	$15a_{1g}[d_{z^2}]$	-11.51	—	—
	$9b_{2g}[d_{xy}]$	-13.01	—	—
	$2e_g\ [d_\pi]$	-14.61	—	—
FeP(Py)	$b_1(e_g)$	0.35	—	—
$[^5B_1]$	$a_2(e_g)$	0.34	—	—
	$b_1(a_{2u})$	-6.65	—	—
	$a_2(a_{1u})$	-6.70	—	—
FeP(Py)(CO)	$a_2(e_g)$	0.23	—	—
$[^1A_1]$	$b_1(e_g)$	0.20	—	—
	$a_2(a_{1u})$	-6.79	—	—
	$b_1(a_{2u})$	-6.86	—	—

The reorganization energy is small, about 0.3∿0.4 eV if an electron is removed from either of the two highest occupied orbitals which are porphine π-orbitals. On the other hand, the reorganization is as big as 6∿7 eV when an electron is removed from a metal d-orbital.

ΔSCF always gives a better agreement with experiment than a simple estimate of -ε. The degree of agreement of ΔSCF values seems to depend on the quality of the basis used. For the two lowest ionization potentials, the differences between observed and ΔSCF values are 0.8∿1.2 eV for the minimal, 0.3∿0.5 eV for the partial double and 0.1 eV or less for the double basis.

Our calculations indicate that the a_{1u} orbital has a higher energy than a_{2u} except in H_2P but the difference is rather small and lies between 0.15 eV and 0.45 eV.

When a two-electron oxidation of the ferrous iron porphyrins occurs in a biological system, the Mössbauer spectra [21] showed that the first oxidation is of the iron. Whether the second oxidation is of the iron or of the porphyrin macrocycle is currently being debated [22]. According to our ΔSCF calculations on ferrous FeP(A), the ionization energy of an electron in d_{z^2} is smaller than that of an electron in the highest occupied π orbital $1a_{1u}$ by 1 eV. This is in accordance with the picture that the first oxidation is of the ion. Similar calculations on FeP^+ would be able to shed a light on the problem whether the second oxidation produce the ferryl complex or the ferric π-cation.

3. EXCITATION ENERGY

The absorptions in the visible and UV regions of metal and metal-free porphines are due to the π electron excitations of the porhine ring.

The first π-CI calculation has been carried out on CoP. From its ground state $^2A_{1g}$, one electron is excited from occupied π-MO's except the $1a_{2u}$ and $2a_{2u}$ orbitals, which are essentially Co 2p and 3p orbitals, to an unoccupied orbitals. This single π-CI yields 164 ^2E_u configuration state functions (CSF's). The excitation energy is calculated as the difference from the SCF energy of the $^2A_{1g}$ state. As is listed in Table 4, the Q peak is too high by 0.8 eV but this is not unexpected from rather crude approximations involved in the calculation. The calculated Soret band is too high by 2.3 eV and this seems to be too big for this type of calculation. It is known that diffuse orbitals sometimes play an important role in some lower excited states of conjugated systems. In fact, in pyrrole, which constitutes a part of porphine, the first excited state has a marked Rydberg character having an extended spacial distribution [25]. A calculation including a diffuse π orbital on each carbon and nitron is thus attempted on CuP at the same level of calculation i.e. π-CI(S). The results given in

Table 4. Excitation energies (oscillator strengths) of Co- and Cu- porphine (eV)

Molecule	CoP	CoOEP	CuP	CuP and CuOEP
Symmetry	^2E_u		^2E_u	
Method / State (Dim.)	π-CI(S) (164)	obs.[23]	π-CI(S) (306)	obs.
T	1.81		1.90(3x10^{-6})	1.88*
Q	3.01	2.24(0∿0.1)	3.05(0.02)	2.25**
B	5.59	3.24(1.1∿1.6)	5.43(7.76)	3.15**
N	5.85	3.84(∿0.1)	5.86(1.27)	3.87***
L	7.65	4.95	—	—

* CuP in n-octane [24]

** CuP in benzene [24]

*** CuOEP [23]

Table 4 show that our conjecture was not correct. Excited states in which diffuse orbitals play an important role have energies higher than 6 eV and the Q, B, and N states are not affected.

The π-CI including the singly and doubly excited configurations with respect to a few reference configurations is tried next, this time on H_2P and H_2C. The CGTO basis is minimal and does not contain diffuse orbitals. The results are given in Table 5. Again there is no significant change over π-CI(S) results as far as excitation energies are concerned. However, the content of wave functions does change when doubly excited configurations are included. The main configurations in the five lowest states of H_2P are shown in Table 6 and compared with the result of Petke et al. [5] whose basis is spherical floating GTO's The 2 $^1B_{2u}$ state which is a main component of the Soret band is worth special attention. At the CI(S) level, the well known Gouterman four orbital model seems to be breaking down but the model recovers, to a great extent, at the πCI(S,D) level. This can be taken as an indication that the nature of Soret band is more complicated than was previously supposed.

Table 5. Excitation energies of free base porphine and chlorine (in eV).

	H_2P		H_2OEP		H_2C		
state	π-CI(S)	π-CI(S,D)	obs.[23]	state	π-CI(S)	π-CI(S,D)	obs.[24a]
$1\ ^1B_{2u}$	2.63	2.27	1.95	$1\ ^1B_1$	2.78	2.46	1.95
$1\ ^1B_{1u}$	3.13	3.42	2.34	$2\ ^1A_1$	3.57	3.64	2.30
$2\ ^1B_{2u}$	4.83	4.72	3.22	$2\ ^1B_1$	5.19	5.00	3.20
$2\ ^1B_{1u}$	5.79	5.29		$3\ ^1A_1$	5.61	5.28	—

The next step to be taken is to include σ-π correlation. Unfortunately it is a tremendous job to include this effect in ab initio CI calculations for a system of this size. The CI program has to be improved first and the task is presently underway.

Table 6. Major configurations resulting from π-CI(S and S,D) calculation of free base porphine

Method / State	This work πCI(S)		This work πCI(S,D)		Petke et al. [5] πCI(S,D)	
$^1A_{1g}$	ground conf.	1.00	ground conf.	0.79	ground conf.	0.86
$1\ ^1B_{2u}$	$5b_{1u}\rightarrow 4b_{3g}$	0.60	$5b_{1u}\rightarrow 4b_{3g}$	0.50	$5b_{1u}\rightarrow 4b_{3g}$	0.48
	$2b_u\rightarrow 4b_{2g}$	0.35	$2a_u\rightarrow 4b_{2g}$	0.26	$2a_u\rightarrow 4b_{2g}$	0.37
$1\ ^1B_{3u}$	$2a_u\rightarrow 4b_{3g}$	0.54	$2a_u\rightarrow 4b_{3g}$	0.45	$2a_u\rightarrow 4b_{3g}$	0.49
	$5b_{1u}\rightarrow 4b_{2g}$	0.44	$5b_{1u}\rightarrow 4b_{2g}$	0.36	$5b_{1u}\rightarrow 4b_{2g}$	0.40
$2\ ^1B_{2u}$	$4b_{1u}\rightarrow 4b_{3g}$	0.54	$2a_u\rightarrow 4b_{2g}$	0.43	$4b_{1u}\rightarrow 4b_{3g}$	0.35
	$2a_u\rightarrow 4b_{2g}$	0.29	$5b_{1u}\rightarrow 4b_{3g}$	0.17	$2a_u\rightarrow 4b_{2g}$	0.30
	$5b_{1u}\rightarrow 4b_{3g}$	0.09	$4b_{1u}\rightarrow 4b_{3g}$	0.17	$5b_{1u}\rightarrow 4b_{3g}$	0.15
$2\ ^1B_{3u}$	$5b_{1u}\rightarrow 4b_{2g}$	0.51	$5b_{1u}\rightarrow 4b_{2g}$	0.42	$5b_{1u}\rightarrow 4b_{2g}$	0.32
	$2a_u\rightarrow 4b_{3g}$	0.39	$2a_u\rightarrow 4b_{3g}$	0.33	$2a_u\rightarrow 4b_{3g}$	0.31
					$4b_{1u}\rightarrow 4b_{2g}$	0.15

4. CHARGE DISTRIBUTION

Mulliken's population analysis, despite of its limitations and ad-hoc assumptions, is probably the most powerful and certainly the simplest tool to extract chemically important information about the charge distribution in a molecule from an LCAO MO wave function. Typical examples of the information obtained are the atomic net charges and the overlap populations between atoms.

In Table 7, the net charge of the metal atom surrounded by the porphine ring with and without 5th and 6th ligands are given. It is remarkable that the net charge lies within the range 1.0∿2.1 irrespective of whether the metal atom is Co, Cu, Fe and independent of its valency (or the formal oxidation number) and spin state. When an electron is removed from the system, at least 50% and sometimes almost 100% of out going charge comes from the porphine ring. This means when an electron is lost from a d orbital of the metal, considerable charge flows from the porphine ring to the $d_{x^2-y^2}$ orbital if it is vacant and also to the 4s and 4p orbitals. Looking through the figures of these tables, the author cannot help being reminded of the Pauling electroneutrality principle. The principle says that each atom in stable complexes has only a small electric charge in the range -1 to +1. This is almost true for net charges in metal porphines calculated in this work (see Table 8).

Overlap populations are often interpreted as an indicator of the strength of a covalent bond between the two atoms involved. Overlap populations between the metal and one of its neighboring atom are collected in Table 9. If the interpretation mentioned above is valid for overlap populations as small as 0.4∿0.05 the covalency between the metal and N seems to increase when the number of electron in the system decreases. A representative example is CoP. There is also a tendency for low-spin states to have a little larger overlap population than high-spin states for a given complex. This is in accord with the general idea that low-spin complexes have more covalency. However, the differences are so small and it is doubtful whether they have really chemical significance.

It is to be noted that FeP has been calculated with two basis, partial double and double. Two sets of calculated net charges and overlap populations agree qualitatively and the maximum difference being 0.17 for the net charge and 0.04 for the overlap population.

5. CONCLUDING REMARKS

Metal porphines are large systems for ab initio computations. Powerful and efficient computer program packages for the LCAO SCF and CI calculations have been developed. Thanks to these programs, ab initio calculations on several metal porphines with up to 300 CGTO's as a basis are being carried out.

Table 7. Calculated net charge of metal in metal porphines

CoP	$^4B_{2g}$	1.86	$^4A_{2g}$	1.88	^4E_g	1.87		
	$^2A_{1g}$	1.78	^2E_g	1.76				
CoP^+	^3E_g	2.06						
CoP^{2+}	$^4A_{2g}$	2.07						
CoP^-	$^1A_{1g}$	1.29						
CuP	$^2B_{1g}$	1.31	$^2A_{1g}$	1.49	^2E_g	1.29	$^2B_{2g}$	1.31
CuP^+	$^1A_{1g}$	1.43	$^1B_{1g}$	1.35	^1E_g	1.32	$^1A_{2g}$	1.32
	^3E_g	1.45	$^3A_{2g}$	1.63				
CuP^-	$^1A_{1g}$	1.01	^1E_g	1.28	^3E_g	1.28		
FeP(A)	$^3A_{2g}$	1.55	$^3B_{2g}$	1.58	$^1A_{1g}$	1.61		
FeP(B)	$^5A_{1g}$	1.49	^5E_g	1.53	$^5B_{2g}$	1.50		
	$^3A_{2g}$	1.39	^3E_g	1.42	$^3B_{2g}$	1.44		
	$^1A_{1g}$	1.44						
FeP(Py)	$^5B_1$	1.50						
FeP(Py)(CO)	$^1A_1$	1.73						

Table 8. Calculated atomic net charges in metal porphines

Molecule	CoP	CuP	FeP		FeP(Py)	FeP(Py)(CO)
State	$^2A_{1g}$	$^2B_{1g}$	$^3A_{2g}$		$^5B_1$	$^1A_1$
			(A)	(B)		
Metal	1.78	1.28	1.55	1.39	1.50	1.73
N	-0.57	-0.42	-0.52	-0.60	-0.46	-0.52
C_α	0.05	0.04	0.05	0.02	0.04, 0.04	0.05, 0.04
C_β	-0.14	-0.15	-0.14	-0.11	-0.14,-0.15	-0.14,-0.15
C_m	-0.12	-0.12	-0.12	-0.18	-0.12,-0.13	-0.12,-0.13
H_β	0.14	0.14	0.15	0.19	0.14, 0.14	0.14, 0.14
H_m	0.14	0.15	0.14	0.23	0.14, 0.14	0.14, 0.14
N					-0.29	-0.37
C_α					-0.03	-0.02
C_β					-0.15	-0.16
C_γ					-0.09	-0.10
H_α					0.17	0.16
H_β					0.15	0.15
H_γ					0.15	0.15
C						0.59
O						-0.56

Table 9. Overlap population between metal and neighboring atoms

CoP	$^2A_{2g}$	0.06						
CoP^+	^3E_g	0.25						
CoP^{2+}	$^4A_{2g}$	0.33						
CoP^-	$^1A_{1g}$	-0.06						
CuP	$^2B_{1g}$	0.24	$^2A_{1g}$	0.24	^2E_g	0.23	$^2B_{2g}$	0.23
CuP^+	$^1A_{1g}$	0.34	$^1B_{1g}$	0.30	^1E_g	0.30	$^1A_{2g}$	0.29
	^3E_g	0.33	$^3A_{2g}$	0.34				
CuP^-	$^1A_{1g}$	0.15	^1E_g	0.24	^3E_g	0.24		
FeP(A)	$^5A_{1g}$	0.02	^5E_g	0.03	$^5B_{2g}$	0.05		
	$^3A_{2g}$	0.08	^3E_g	0.09	$^3B_{2g}$	0.07		
	$^1A_{1g}$	0.09						
FeP(B)	$^5A_{1g}$	0.05	^5E_g	0.06	$^5B_{2g}$	0.08		
	$^3A_{2g}$	0.12	^3E_g	0.13	$^3B_{2g}$	0.12		
	$^1A_{1g}$	0.12						
FeP(Py)	$^5B_1$		FeN(P)	0.15	FeN(Py)	0.06		
FeP(Py)(CO)	$^1A_1$		FeN(P)	-0.01	FeN(Py)	-0.37	FeC	-0.12

Ionization potentials of porphine π-electrons calculated by the Δ-SCF method with a basis set of "partial double" agree with experiment within an error of about 0.3 eV. A similar calculation with d-electron ionization will give useful information about oxidation of the metal porphines.

The calculated Soret band by singly excitation π-CI is too high by 2.3 eV and no improvement is made by including diffuse orbitals in a basis nor by adding double excitations with respect to a few reference configurations in the π-CI. The σ-π correlation may be important in correcting this error.

Mulliken's population is used to obtain ideas about the charge distribution in the molecules. The net charge on the metal is in the range of 1.0∼2.1 irrespective of whether it is Co, Cu or Fe and independent of its valency and the spin state.

The work as a whole is still at its initial stage and there are high hopes for broadneing it to include many properties and various spectroscopic data.

The electronic structure of these biologically important molecules will be understood better by an interplay between experiment and two kinds of theories namely semi-empirical and ab-initio. Ab initio studies are worth the effort and this will be made by both development of formal theories and innovations in data handling on a computer.

APPENDIX 1. "JAMOL3", AN LCAO SCF PROGRAM

An LCAO SCF program "JAMOL3" is an improved version of "JAMOL2" and has been written by H. Kashiwagi, T. Takada, E. Miyoshi, S. Obara and F. Sasaki.

This program can deal with closed- and open-shell systems in the manner of Roothaan [26]. Only one open shell for each symmetry is permitted. A basis can consists of more than 300 CGTO's having s, p, d, and f characters.

The program makes use of symmetry in various steps, especially in reducing a number of two electron integrals (TEI's) to be evaluated. The symmetries considered involve C_1, C_s, C_i, C_2, D_2, D_4, C_{2v}, C_{4v}, C_{2h}, D_{2h}, D_{4h}, D_{2d} and O_h. The number of necessary TEI's is given roughly by the total number ($\sim N^4/8$ where N is the number of CGTO's) divided by the number of symmetry operation. For example, when 184 CGTO's are used for CoP with D_{4h} symmetry, the total number of TEI's is 155×10^6 among which the linearly independent TEI's are 10×10^6.

A further reduction of the number of integrals is achieved by using an integral approximation scheme [27,28,29] based on semi-orthogonalized orbitals [30]. A generalized Löwdin orbital set g is characterized by the following three requirements.

(i) The overlap matrix of g is prefixed to an arbitrarily assigned positive definite matrix T.

(ii) The set g is a linear transform of the original orbital set f.

(iii) The set g minimizes the sum

$$\sum_i \int |g_i - f_i|^2 dv \ .$$

The semi-orthogonalized orbitals are defined by the following choice of the matrix T. If orbitals f_i and f_j have an overlap integral S_{ij} whose absolute value is less than a threshold δ_1 and the exchange integral $\langle ij|ji\rangle^f$ is less than another threshold δ_2, the pair (i, j) is called a weakly related pair and T_{ij} is set to zero. If the pair (i, j) is not weakly related, T_{ij} is set to the original overlap integral S_{ij}. The integral evaluation scheme is as follows:

(i) All one electron integrals in the g basis are evaluated rigorously from those in the f basis.

(ii) A TEI $\langle ij|kl\rangle^g$ is neglected if both or one of the orbital pairs (i, j) and (k, l) are weakly related. Otherwise, $\langle ij|kl\rangle^g$ is replaced by $\langle ij|kl\rangle^f$.

This approximation (ii) cuts the number of TEI's to be calculated to the order of N^2 without loosing too much accuracy if the number of orbitals per atom and the thresholds δ_1 and δ_2 are fixed.

There are two more options in "JAMOL3" concerning integral approximations.

The first is to evaluate a contribution from TEI over primitive GTO's to a TEI over SO's and if the contribution is less than the third threshold δ_3, the TEI over primitive GTO's is not calculated and is neglected.

The second is to neglect all integrals whose absolute values are smaller than the fourth threshold δ_4.

When a molecule under investigation is as large as metal porphines, an integral approximation scheme such as explained above is quite effectual and seems almost necessary.

In "JAMOL3", TEI's are evaluated by a program called "GINT2" which is developed by F. Sasaki. TEI's over GTO's have a form

$$I(f_1, f_2) = \int dv \int dv' \frac{1}{|r-r'|} f_1(r) f_2(r') e^{-\alpha_1 (r-R)^2 - \alpha_2 (r'-R')^2}$$

where f_1 and f_2 are polynomials of x, y, z which are determined by the orbital pairs (i, j) and (k, l), respectively. In "GINT", the formula

$$I(f_1, f_2) = \sum_k b_k \frac{\partial^k}{\partial(s^2)^k} F_o(\mu s^2) = \sum_k C_k F_k(\mu s^2)$$

is used, where

$$\mu = \frac{\alpha_1\alpha_2}{\alpha_1+\alpha_2} ,$$

$$S = R - R' ,$$

$$F_k(y) = \int_o^1 t^{2k} e^{-yt^2} dt = (-1)^k \frac{d^k}{dy^k} F_o(y) ,$$

and b_k and c_k are suitable constants. The present version of "GINT2" is very efficient.

The most time-consuming part of the LCAO SCF calculation turned out to be the transformation of TEI's from the AO basis (in case of "JAMOL3", CGTO's) to those of the SO's. An improved algorithm [31] over the one used in "JAMOL2" for this process has been incorporated. When the system under consideration has high symmetry such as D_{4h}, the new algorithm requires a number of multiplications proportional to only the fourth power of a number of CGTO's in an SO instead of the usual fifth power. The CPU time needed for transformation has been reduced by a factor 3 by this algorithm in the case of CuP with D_{4h} symmetry.

When using a large basis, one has to be careful about approximate linear dependence. This is particularly true for a large system. In JAMOL3, the overlap matrix is diagonalized and if the smaller diagonal matrix elements are smaller than another threshold, corresponding orbitals may be excluded from the calculation. Timing data of the LCAO SCF calculations reported in this paper are collected in Table 10.

APPENDIX 2. "COMICAL2", A CI PROGRAM

This program has been written by K. Tanaka, H. Tatewaki, T. Nomura and T. Noro and is an improved version of "COMICAL1". Some of its characteristic features are as follows.

For a particular state of interest, the program can have multiple reference configurations and generates configuration state functions (CSF's) by exciting one and two electrons from each reference configuration. A restriction on CSF's generated by "COMICAL2" is that it is space symmetry adapted only when the symmetry group is Abelian.

Table 10. Timing data of the LCAO SCF calculation

Molecule	Symmetry	Prog. JAMOL	Calculated TEI (in 10^6)	Integral Approx. δ_1	δ_2	δ_3	δ_4	Integrals & Supermatrix (in minutes)	Each Iteration (in minutes)	Computer*
H_2P	D_{2h}	3	0.6	$2x10^{-3}$	$3x10^{-5}$	0	0	86	0.2	M-180
H_2C	C_{2v}	3	1.3	$2x10^{-3}$	$3x10^{-5}$	0	0	169	0.5	M-180
CoP	D_{4h}	2	2	$7x10^{-4}$	0	0	0	630	1	230-75
CuP	D_{4h}	2	0.9	$2x10^{-3}$	$3x10^{-5}$	0	0	211	0.5	230-75
FeP(A)	D_{4h}	3	2.6	$1x10^{-3}$	$1x10^{-6}$	$1x10^{-6}$	$1x10^{-6}$	120	0.3	M-180
FeP(B)	D_{4h}	3	8	$1x10^{-3}$	$1x10^{-5}$	0	0	546	2	M-180
FeP(Py)	C_{2v}	3	11.5	$1x10^{-3}$	$1x10^{-6}$	$1x10^{-6}$	$1x10^{-6}$	394	1.4	M-180
FeP(Py)	C_{2v}	3	17	$1x10^{-3}$	$1x10^{-6}$	$1x10^{-6}$	$1x10^{-6}$	547	2	M-180

* The CPU speed of FACOM 230-75 and HITAC M-180 are very roughly 1.1 and 1.5 times that of IBM 370-168.

CSF's are expressed in terms of holes and particles with reference to an assumed closed shell. Expression for the Hamiltonian matrix elements are written by using the Fock matrix of the closed shell and TEI's. This eliminates the need of full transformation of the TEI's and also helps to reduce greatly the number of terms stored in a code tape for the Hamiltonian matrix.

Further selection of CSF's is, in most cases, necessary. This is made first by constructing the first order interacting space in a manner developed by Sasaki which is described elsewhere [32]. A perturbative selection option is also offered in the program. The method is essentially Shavitt's B_k method [33]. The whole space is divided into two, A and B, where the A space is spanned by the reference functions. The matrix equation to be solved is

$$H_A^{eff} C_A = EC_A$$

where

$$H_{AiAj}^{eff} = H_{AiAj} + \sum_{\alpha}^{(B)} \frac{H_{AiB\alpha} H_{B\alpha Aj}}{E - H_{B\alpha B\alpha}} .$$

This equation is solved by the Newton-Raphson procedure. An estimate of the contribution to the ith root of H_A^{eff} from a CSF α in the B space $\Delta E_\alpha^{(i)}$ is given by

$$\Delta E_\alpha^{(i)} = \langle C_A^{(i)} | \frac{H_{AiB\alpha} H_{B\alpha Ai}}{E - H_{B\alpha B\alpha}} | C_A^{(i)} \rangle .$$

When $|\Delta E_\alpha^{(i)}|$ is smaller than a threshold δ for any root i of interest, the CSF α is thrown away. By summing up $\Delta E_\alpha^{(i)}$ over the thrown away CSF's, $\Delta E_{thrown}^{(i)}$ is defined;

$$\Delta E_{thrown}^{(i)} = \sum_{\alpha}^{(thrown)} \Delta E_\alpha^{(i)} .$$

The ith solution of the CI involving selected CSF's $E^{(i)}$ and also $E^{(i)} + \Delta E_{thrown}^{(i)}$ can be extrapolated to δ=o when a series of calculations are done with a few thresholds. This extrapolation works better when it is made against E_{thrown} rather than the threshold itself, as was noted by Shavitt [34].

An natural orbital iteration can be made automatically in the program.

Some timing data for the planar form of formylmethylene and pyrrole are given Table 11.

Table 11. Timing data of CI calculation

Molecule	HCCHO	Pyrrole
Symmetry	$^1A''$	$^1A_1$
No. of MO's	28	64
No. of CSF's generated	9062	57,958
No. of CSF's selected	2306	3,185
δ(μH)	10	50
CPU time (in minute on a HITAC M-180 computer)		
interface with JAMOL3	0.4	5
transformation of TEI	0.6	6
CSF generation	2.2	39
perturbative selection	3.6	18
H-matrix expression	33.0	45
H-matrix evaluation	1.5	5.5
diagonalization	1.5(one root)	8(three roots)
Total	42.8	126.5

ACKNOWLEDGMENT

All the work reported in this paper has been carried out by my collaborators Prof. H. Kashiwagi, Mr. S. Obara, Dr. T. Takada, Dr. E. Miyoshi, Mr. U. Nagashima, Dr. K. Tanaka, Dr. H. Tatewaki, Mr. T. Nomura, Dr. T. Noro and, last but not least, Prof. F. Sasaki. The author wishes to express his hearty thanks and total indebtedness to these collaborators. He is also very grateful to Prof. H. S. Taylor for his linguistic help.

REFERENCES

[1] Almlöf, J.: 1974, Int. J. Quant. Chem. 8, 915.
[2] Spangler, D., Mckinney, R., Christoffersen, R. E., Maggiora, G. M., and Shipman, L. L.: 1975, Chem. Phys. Lett. 36, 427.
[3] Spangler, D., Maggiora, G. M., Shipman, L. L., and Christoffersen, R. E.: 1977, J. Am. Chem. Soc. 99, 7470.
[4] Spangler, D., Maggiora, G. M., Shipman, L. L., and Christoffersen, R. E.: 1977, J. Am. Chem. Soc. 99, 7478.
[5] Petke, J. D., Maggiora, G. M., Shipman, L. L., and Christoffersen, R. E.: 1978, J. Molec. Spectry. 71, 64.
[6] Dedieu, A. and Rohmer, M. M.: 1977, J. Am. Chem. Soc. 99, 8050.
[7] Rohmer, M. M., Barry, M., Dedieu, A., and Veillard, A.: 1977, Int. J. Quant. Chem. Quant. Biol. Symp., 4, 337.
[8] Nagashima, U., Takada, T., Tanaka, K., and Ohno, K.: unpublished.
[9] Kashiwagi, H., Takada, T., Obara, S., Miyoshi, E., and Ohno, K.: 1978, Int. J. Quant. Chem. 14, 13.
[10] Obara, S., Kashiwagi, H., Takada, T., Miyoshi, E., Nagashima, U., Sasaki, F., and Ohno, K.: unpublished.
[11] Obara, S. and Kashiwagi, H.: unpublished.
[12] Kashiwagi, H. and Obara, S.: unpublished.
[13] Obara, S. and Kashiwagi, H.: unpublished.
[14] Obara, S. and Kashiwagi, H.: unpublished.
[15] Roos, B. and Siegbahn, P.: 1970, Theor. chim. Acta 17, 209.
[16] Huzinaga, S.: 1965, J. Chem. Phys. 42, 1293.
[17] Roos, B., Veillard, A., and Vinot, G.: 1971, Theor. chim. Acta 20, 1.
[18] Huzinaga, S.: 1977, J. Chem. Phys. 66, 4245.
[19] Van Duijneveldt, F. B.: 1971, I.B.M. Tech. Res. Rept. No. RJ945.
[20] Kitagawa, S. and Morishima, I.: private communication.
[21] Moss, T. H., Ehrenberg, A., and Bearden, A. J.: 1969, Biochemistry 8, 4159.
[22] Dolphin, D., Addison, A. W., Cairns, M., Dinello, R. K., Farrell, N. P., James, B. R., Paulson, R. R., and Welborn, C.: 1979, Int. J. Quant. Chem. 14, 311.
[23] Edwards, L., Dolphin, D., and Gouterman, M.: 1970, J. Mol. Spectry. 35, 90.
[24] Eastwood, D. and Gouterman, M.: 1969, J. Mol. Spectry. 30, 437.
[24a] Eisner, U. and Linstead, R. P.: 1955, J. Chem. Soc. 3742.
[25] Tanaka, K., Nomura, T., Noro, T., Tatewaki, H., Takada, T., Kashiwagi, H., Sasaki, F. and Ohno, K.: 1977, J. Chem. Phys. 67, 5738.
[26] Roothaan, C. C. J.: 1960, Rev. Mod. Phys. 32, 179.
[27] Kashiwagi, H.: 1976, Int. J. Quant. Chem. 10, 135.
[28] Kashiwagi, H. and Takada, T.: 1977, Int. J. Quant. Chem. 12, 449.
[29] Osanai, Y. and Kashiwagi, H.: accepted for publication in Int. J. Quant. Chem.
[30] Kashiwagi, H. and Sasaki, F.: 1973, Int. J. Quant. Chem. S7, 515.
[31] Takada, T. and Sasaki, F.: submitted for publication in Int. J. Quant. Chem.

[32] Tatewaki, H., Tanaka, K., Sasaki, F., Obara, S., Ohno, K., and Yoshimine, M.: 1979, Int. J. Quant. Chem. 15, 533.
[33] Gershgorn, Z. and Shavitt, I.: 1968, Int. J. Quant. Chem. 2, 751.
[34] Shavitt, I.: unpublished.

THEORETICAL AND COMPUTATIONAL CHEMISTRY OF COMPLEX SYSTEMS

Enrico Clementi
IBM DPPG
P.O. Box 390, B28, 702-2
Poughkeepsie, N.Y. 12602, U.S.A.

ABSTRACT We present a general operational method aimed at quantitative simulations of complex chemical systems. Quantum-chemical simulations are used only for those aspects of the problem where electrons and nuclei must be explicitly considered as the system's particles. Statistical mechanics is used for those aspects where atoms and molecules can be considered as the system's particle; other aspects are simulated assuming a continuoum distribution rather than a discrete distribution. The parameters in the general method are internal energy, electronic density, entropy, temperature, free energy, probability distribution, time and static or dynamic properties of continuoum media. Examples considered deal with solvent-solute interactions where the solute are either ions, or amino acids, or proteins, or bases and pair-base of nucleic acids, or DNA (single or double helix).

INTRODUCTION

The complexity of a chemical system increases proportionally to the number of non-independent variables needed to describe a given aspect of the system. There are many different possible ways to classify the non-independent variables; to classify the degree of complexity of a chemical system we select to group the variables into those related to the description of the largest molecule of the system (size as criterium) and into those related to the interaction between molecules, using the number of molecules in the system as second criterium. According to the above definition a chemical system is considered to be complex either because it contains a macromolecule or because it contains many molecules or both; "size" (defined as the number of atoms

K. Fukui and B. Pullman (eds.), Horizons of Quantum Chemistry, 267–279.

constituting the largest molecules) and "number" can be considered two orthogonal axis defining the "complexity" plane. A system composed of liquid water compared to a system of one water molecule constitutes an example of increasing complexity (from high to low) due to the "number" criterium; equivalently, a system composed of a protein conpared to a system containing one amino acid constitutes an example of increasing complexity due to the "size".

The most basic selection one makes to describe a system with some given model, relates to the choice of the adopted statistic; this choice brings about a specific definition of the nature of the "objects" (particles or medium) and concomitant model-equations. "Natural objects" for chemical systems are either nuclei and electrons or atoms and molecules or some continuous medium (rigid or non-rigid). The corresponding statistics are either the Fermi-Dirac statistic or the classical statistic (inclusive of Boltzmann distribution); the corresponding equations are either the Schröedinger equation or Newton's equations (for discrete particles or for continuous medium). In the latter case of particular chemical interest are aspects dealing with the thermodynamic of reversible and irreversible systems, problems of linearity and non-linearity, single or multiple solution for stationary states, etc.

Today, quantum chemistry has sufficently evolved so as to allow us to consider realistically complex systems as an object for numerical simulations; as below described, however, it is basic to reject the indiscriminate use of quantum chemistry as the only tool to be used in complex system's description. Such indiscriminate use brings about necessarily gross over-simplifications, that can be avoided by using quantum-chemistry as a first step of a many step methodology aimed at realistic simulations of complex chemical systems.

A GENERAL METHOD

We briefly summarize an operational procedure to simulate complex chemical systems. Most of the theoretical foundations needed in our approach have been know since long ago; however little has been done to operationally link different methods, that are traditionally kept independent one from the other, despite the fact they represent the successive step of our approach.

Let us start by considering some appropriate subsystem of our complex system. The simplest and most immediate subsystems are the individual, separated molecules. These --

due to the size -- are either amenable to quantum chemical computations or not. As known, the size of a molecule for which "decent" quantum chemical simulations are feasible is constantly increasing either because of new methods or because of the increased performance of computers. "Decent" simulations are those that make use of adequate basis sets and that include electronic correlation corrections, when needed. In this context it is of interest to point out a few recent findings concerning quantum chemical computations. It has been proved(1) that the use of minimal basis sets (not sub-minimal basis sets as the STO-3G) plus the use of the counterpoise method (2) yield inter-molecular interaction energies nearly as accurate as those obtained with extended (Hartre-Fock type) basis sets. In addition, we have found that dispersion corrections can be reliably added to Hartre-Fock type inter-molecular interactions (as perturbation) to account for the most important part of the electronic correlation correction.(3) Finally, we have carried out computations, proving that the non-additive part of three-body corrections can be cast into the form of induction interactions.(4) As a result of the findings, two-body and three-body (and an important part of the n-body) interactions needed to represent intermolecular potentials can now be easily obtained; as known, prior to this work, the n-body corrections were somewhat like a nightmare for quantitative simulations.

There are, however, single molecules of such large size as to render most unpractical any direct quantum-chemical computation. However, in the case of such macromolecules the main interest in quantum-chemistry is related to solutions of two types of problems, a) conformational problems and b) reaction pathway of the macromolecule with a second molecule. Recently we have proposed and tested methods that are of ab-initio type and can deal with the above problems. For example, we have considered the conformational problem for nucleic acids(5) and the energy of reaction for the system composed by a substrate interacting with the enzyme papaine.(6) Semi-empirical techniques have been proposed and used; however in our general method we have selected to constrain ourself to ab-initio computations at least for the first order approximation (equivalent to the Hartre-Fock model).

With the above recent advances in the methodology for systems of intermediate complexity, it is now feasible to proceed to a brief outline of a general operational method aimed not only at simple and intermediate, but also at complex chemical systems. The general method is presented as a five-step technique; the output of step (i-1) constitutes the input to step (i). This constrain is basic to develop an operational procedure.

STEP 1: QUANTUM-CHEMISTRY

Ab-initio quantum chemical simulations are performed on one or few molecules. For a single molecule we are interested in the net atomic charge(7) and in the Molecular Orbital Valence State energy.(8) However, if the separated molecule is a macromolecule then the above quantities can be obtained from fragments (of proper size) of the macromolecule, with or without addition of the field of the rest of the molecule as point charge external field perturbation. It is noted that ab-initio computations with molecules of more than 50 atoms (with Z between 1 and 18) are feasible, but expensive, thus a proper fragmentation can be used to save computer time. If the ab-initio computation is performed in order to obtain inter-molecular interactions, then we proceed at first by obtaining the two-body interaction at the minimal basis set level (e.g. 7s gaussians and 3 to 4 p gaussians for $1s^2 2s^2 2p^n$ atoms); then we correct for the superposition error with the counterpose method.(2) When this is done, the two-body interaction is corrected by obtaining, via perturbation, the dispersion energy. If three-body are considered the same procedure is used, but we add the non-additive part by obtaining the induction energy correction. The number of ab-initio computations to be performed in order to obtain accurate two- and three-body potentials can be initially rather large (for the potential library thus far obtained about five thousands ab-initio computations have been performed).

STEP 2: CONSTRUCTION OF INTERACTION POTENTIALS

From the numerical potentials of Step 1, we obtain analytical potentials. This step constitutes a most critical aspect to operationally connect quantum chemistry to statistical mechanics. The potentials must be constrained in such a way as to be of analytically simple form, fast for computational use, tranferable from molecule to molecule, standardized both in the form and reliability and, finally, amenable to gradual refinements and extensions. By design we have neglected the possibility to use experimental data as parameters of our potentials. One reason is that often there are not sufficent data available; a second reason is that it is often arbitrary to extract two- and three-body contributions from an experiment obtained at conditions corresponding to a full n-body potential.

Basic to this step it is an analysis of how an atom can be characterized when in a molecule. From the valence concepts of Lewis and Langmuir, the valence bond approximation was derived by Heitler, London, Slater, and Pauling; an important concept in that language is the Valence State concept. Alternatively, from the Lewis-Langmuir concepts we can arrive to the Mulliken-Hartre-Fock Molecular Orbitals approximation. Eristically, an important concept connected to the M.O. model is that partitioning of the electronic density known as the electron population analysis. From the Valence Bond approximation we have translated into the M.O. theory an extended version of the concept of Valence State and called it Molecular Orbital Valence State, MOVS. An atom in a molecule can be characterized by its value of the MOVS energy and by the value of its net charge: hybridization, charge transfer and nearest neighbors are included in this characterization. By definition we characterize an atom of a given atomic number and in a specific electronic environment with a label referred to as "class" for that atomic specification. In this way, atoms of the same atomic number and in the same molecular environment belong to the same class. Hence, by construction, the identification of a "class" for an atom is transferable from molecule to molecule, as long as the atom has the same Z value and class label. As a corollary, atom-atom pair potential for atoms of given classes are transferable from molecule to molecule. Historically, we can note the following evolution in the characterization of atoms in molecules: the first and gross characterization is the one corresponding to the atomic number, (e.g. for carbon atoms Z = 6); then the hybridization characterization (e.g. sp, sp^2 and sp^3) and the valence concept; now the "class" characterization is added, providing a finer characterization. For example the carbon atoms in CH_4, CH_3-CH_3, CH_3-NH_2, CH_3-OH, CH_3-SH, have all the same valence, but each one belongs to a different class, as clear from the different values of the two parameter MOVS energy and net charge.

Operationally the transferability of the pair-potentials can be defined as follows: ability to reproduce the intermolecular interaction energy (obtained from ab-initio computations at sampled positions on the energy hypersurface) within a pre-defined threshold error (generally smaller than three to five percents).

At present a library of atom-atom pair potentials has been constructed for pairs composed of the H, C, N, O, P, S, Na^+ and Li^+ atoms. A second library has been constructed for ions interacting with water. The extension of such a library to atoms or classes not presently considered is becoming an increasingly easier task, since more and more a significant number of the atoms constituting the molecule belong to classes analyzed and included into the pair-potential library. The presently available library can be used to obtain the interaction energy between water, selected ions (Li^+, Na^+, K^+, F^-, Cl^-, Be^{++}, Mg^{++}, Ca^{++}, Zn^{++}), any naturally occurring amino acid in either the neutral or zwitter-ion or polar forms, the DNA bases, the sugar and the phosplate units. Among the problems that can now be realistically simulated we mention the structure of liquid water and water clusters, with or without ions, solvation of proteins, enzymes and nucleic acids, study on molten salts, ions effect on enzymes, conformational transition in nucleic acids induced by different ions in solution, determination of the tertiary structure of proteins, interaction between nucleic acids and amino acids or proteins. None of this could be amenable of quantitavie simulations without an atom-atom pair potential library.

STEP 3: STATIC PROPERTIES

The availability of atom-atom pair potential allows us to easily pass to statistical thermodinamics. The static properties are first analyzed. The basic tool in this step is the Monte Carlo method(9) that allows us to introduce temperature averaging in the complex system. We generally use the method at constant volume with periodic boundary conditions and at constant pressure for clusters studies. The introduction of temperature eliminates the un-physical (but currently used) approximation to study water solutions at zero temperature. As known, the reactivity of a system is related to its free energy and not to the internal energy. In this regard we note that the simulation of the entropy is now becoming more feasible: since the Monte Carlo technique cannot yield the entropy because of numerical-statistical problems, the entropy can be obtained by the interaction of the real system (analyzed by the Monte Carlo method) with a fictitious system of ghost molecules that do not evolve on the Markoff chain and do not interact with each other. This method, previously applies to obtain the entropy in the gases of Br_2 and I_2 is now used to obtain the entropy in water solution.(10) Another approach has been recently use.(11)

Recently the Born-Onsager approximation has been reanalized(12,13) mainly to include fully the expansion of electrostatic potentials. We have considered the possibility to use the Born-Onsager approximation mainly to reduce the computational cost of the M.C. technique. To obtain this we have included in Born-Onsager cavity the first two-solvation shells (improperly analyzed in the Born-Onsager approximation) and as many molecules of water as needed to fill a sphere containing the solute and the two shells. The method has been used in simulations of glycyne (neutral and zwitter-ion) and serine (neutral and zwitter-ion). The polarization correction including terms up to the octupole (included) brings about a correction of about 15 to 20 KJ/mol to be compared with a total solvent-solute interaction of 5000 KJ/mol.

STEP 4: DYNAMICAL PROPERTIES

In this step the time parameter is introduced in the standard form of molecular dynamics. Prerequisite for this step is the availability of pair-potentials from Step 1. As initial condition to describe the system we use the final configuration obtained from Step 3. Work is in progress on this step; we note, however, that from the current and past literature on applications in M.D. one can obtain a realistic estimate on the importance of this step. The transport aspect is one of the main result that can be obtained from the time parameter introduction.

STEP 5: CONTINUOUM PRESENTATION

In this step, the basic coefficients needed to solve, for example the diffusion equations of a flow are obtained from Steps 3 and 4. As in the case of Step 4, we are only beginning to work at this problem, but the very ample literature in fluido-dynamics coupled with the ample literature on biological dissipative systems, and time or/and space fluctuations should be enough to let one understand how much rewarding will be to unify the scope of Step 5 with the outputs of Steps 3 and 4. Traditionally this step is the less operationally connected to quantum mechanics and quantum chemistry.

APPLICATIONS

Space limitations do not allow to discuss in any detail the numerous applications thus far obtained. However, extensive details on Step 2 can be obtained from a number of papers.(14) From the early application of Step 3 concerning liquid water(15) and liquid water with ions,(16) we have progressed to solutions containing amino acids or an enzyme(17) as solute.(18) The active site of an enzyme has been also analyzed and the finding on the number of ligands, including water molecules, around the Zn^{++} ion in the carbonic anhydrase has allowed a realistic study of the reaction pathway.(19)

In the studies concerning nucleic acids we have previously reported on the phosphate group(20) on the sugar and on the A-and B-DNA in the single helix.(21,22) More recently, we have considered a fragment of B-DNA double helix, composed of thirteen base-pairs and of the corresponding sugar and phosphate groups for a total of 760 atoms and thousands of electrons. Clearly no reliable quantum mechanical computation can be performed with this fragment. A Monte Carlo simulation has been performed adding 260 water molecules to the above double helix fragment; the water molecules have been constrained in a cylinder of 14.5Å radius and 40.0Å length. This cylinder is co-axial to the DNA cylinder; the DNA fragment analyzed extends to 42.0Å in length and its radius, from the center to the phosphorous atoms, is 8.75Å.

In Figure 1 we report 10 base-pairs considered in the

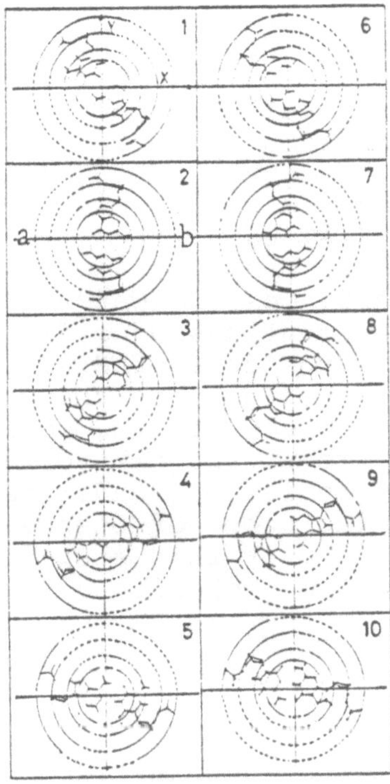

Figure 1. B-DNA fragment used in the Monte Carlo simulation; the inserts 1 to 10 correspond to the sequence in the fragment. The dotted lines refer to hydrated regions, the full lines to hydrophobic regions.

DNA fragment; we have not reported, however, the first two and last base-pair, because of the different boundary conditions of the two ends of the DNA fragment. The circumference given as dotted lines indicates attraction to water (thus water penetration), the solid line portion of the circumference indicates repulsion to water (the quantitative study is reported elsewhere(23)). In Figure 2, we present

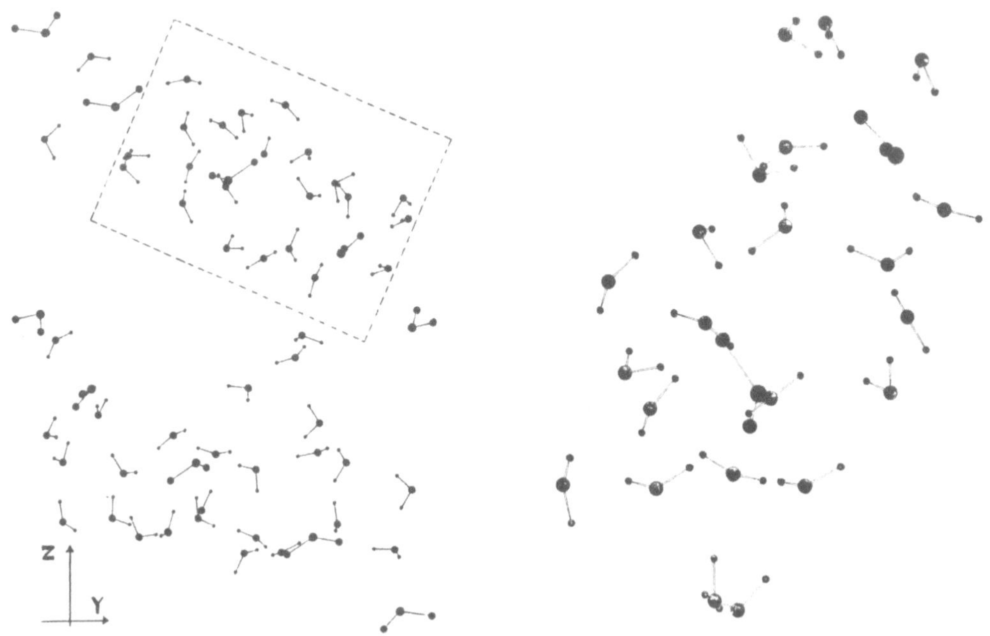

Figure 2. Orientation of a subset of about fifty water molecules around six phosphate groups in the B-DNA fragment (see Figure 1). At the left, a finer detail is given, considering only two phosphate groups.

a detail of the water structure (obtained by Monte Carlo simulations) around few phosphate groups of the DNA fragment. the water structure is, as expected different from the one found around the separated bases (see Figure 3, for the oxygen atom density maps at 300 K and at 200 K) or around the base-pairs (see reference (24)). The water structure is also

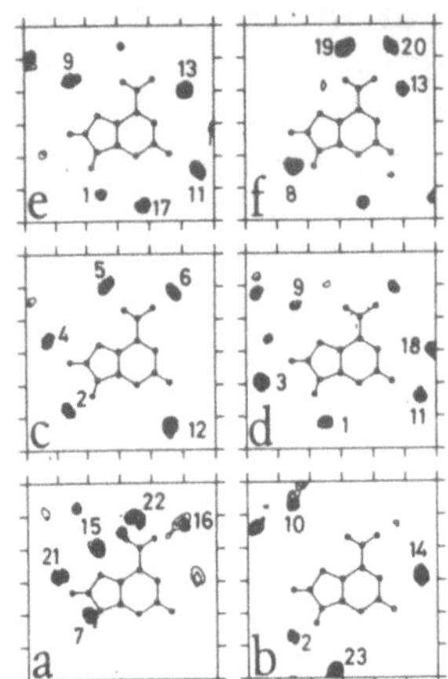

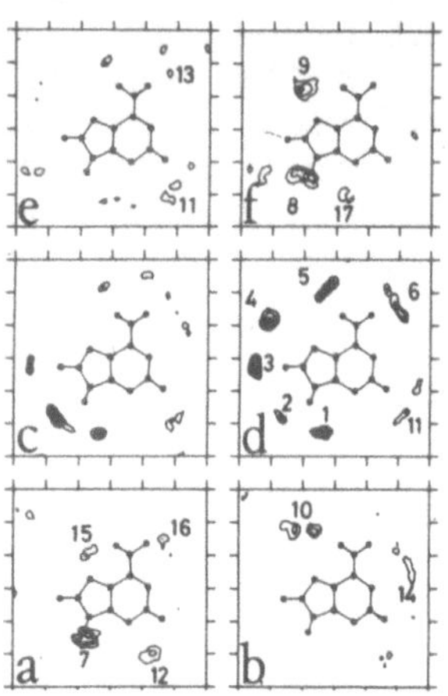

Figure 3. Monte Carlo probability density maps for water's oxygen atoms around adenine at T = 300 K (right) and at T = 200 K (left). Each insert, a to f, represents contiguous volumes above the molecular plane (a,b,c,) and below (d,e,f).

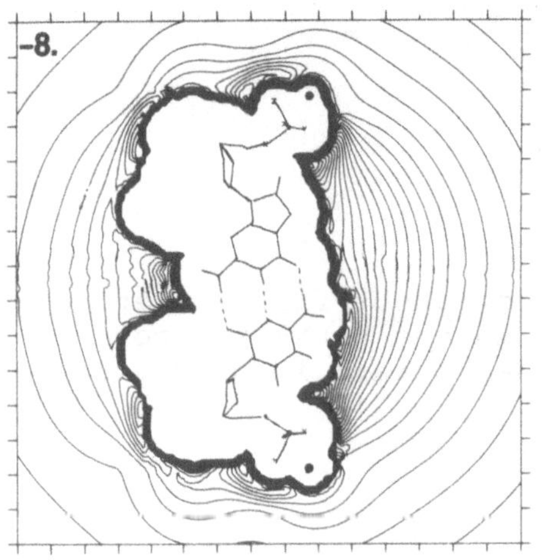

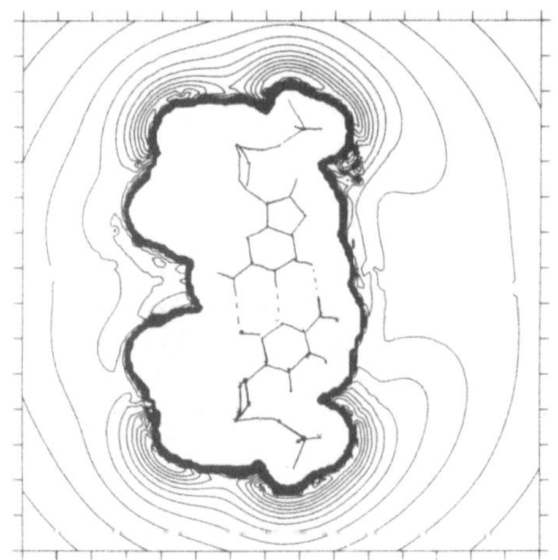

Figure 4. Iso-energy contours for a water molecule interacting with a base-pair (insert 2 of Figure 1) neutralized by Na^+ ions (right) or not (left).

different from what was found when an Na^+ ion neutralizes the PO_4^- groups as clearly visible from Figure 4. All the above computations are obtained with two-body pair potentials. In Figure 5 we report the effect of inclusion of three and n-body corrections.

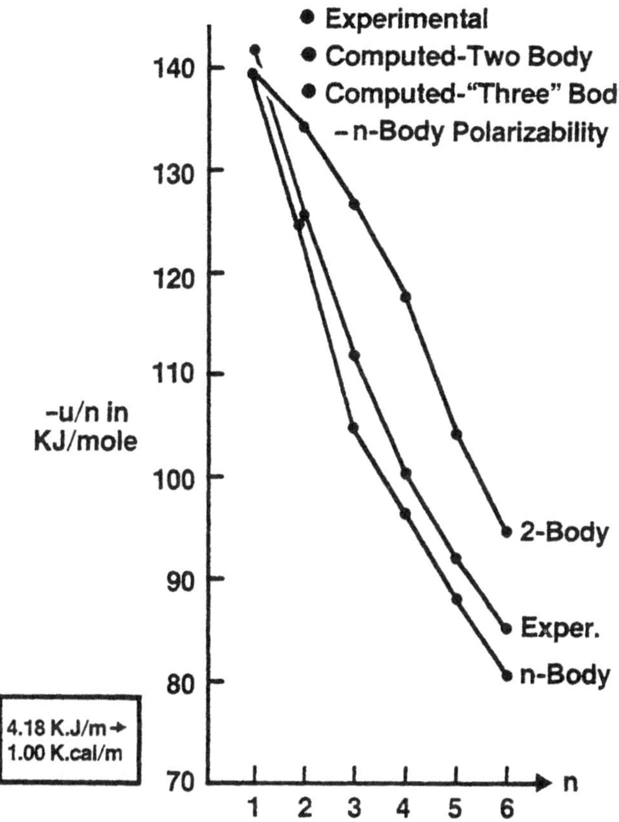

Figure 5. Monte Carlo simulation of $Li(H_2O)_n^+$ water clusters with two- and with n-body potentials at T = 300 K.

These very brief indications are given only as a short illustration of the remarkable possibilities implied in the use of our proposed general method for the quantitative study of complex systems. At the moment our vocabulary is still limited to interaction energy, electronic density, reaction field, temperature, entropy, probability distributions and pair correlation functions. Work is in progress to extend it along the directions outlined in the previous sections.

REFERENCES

1. W. Kolos, Theor. Chim. Acta, (1979, in press).
2. F. S. Boys and F. Bernardi, Mol. Phys., 19, 553 (1970).
3. W. Kolos, E. Clementi, G. Ranghino, S. Romano, Theor. Chim. Acta. (in press).
W. Kolos, G. Corongiu, E. Clementi, J. Inter. Quant. Chem. (in press).
4. E. Clementi, H. Kistenmaker, W. Kolos, S. Romano, Theor. Chim. Acta, (in press).
W. Kolos, G. Lie, G. Ranghino, E. Clementi, J. Inter. Quant. Chem., (in press).
5. O. Motsuoka, C. Tosi, E. Clementi, Biopolymers 17, 33 (1978); 17, 51 (1978); 17, 67 (1978; 18, 203 (1978).
6. G. Bolis, E. Clementi, M. Ragazzi, D. Salvaderi, D. R. Ferro, J. Inter. Quant. Chem. 14, 815 (1978).
7. R. S. Mulliken, J. Chem. Phys., 23, 1833 (1955); 23, 1841 (1955); 23, 2338 (1955).
8. E. Clementi, J. Chem. Phys. 46, 3842 (1967): E. Clementi, A. Routh, J. Inter. Quant. Chem., 6, 525 (1972).; G. Corongiu, E. Clementi, Gazz. Chim. It. 108(5-6), 273 (1978).
9. N. Metropolis, A. W. Rosenbluth, M. N. Rosenbluth, A. H. Teller, E. Teller J. Chem. Phys., 21, 1087 (1973).
10. S. Romano, K. Singer; Mol. Phys., 37, 1765 (1979).
11. S. Swaminathan, D. L. Beveridge, J. Am. Chem. Soc., 99, 8392 (1977).
12. H. L. Friedman, Mol. Phys. 29, 1533 (1975).
13. G. Hall.
14. E. Clementi, H. Popkie, J. Chem. Phys. 57, 1077 (1972); E. Clementi, H. Popkie, H. Kistenmacher, J. Chem. Phys., 58, 1689 (1973); H. Kistenmacher, H. Popkie, E. Clementi, J. Chem. Phys., 58, 5627 (1973); H. Popkie, H. Kistenmacher, E. Clementi, J. Chem. Phys. 59, 1325 (1973); O. Matsuoka, M. Yoshimine, E. Clementi, J. Chem. Phys. 64, 1351 (1976); E. Clementi, F. Cavallone, R. Scordamaglia, J. Am. Chem. Soc., 33, 5531 (1977); R. Scordamaglia, E. Clementi, F. Cavallone, J. Am. Chem. Soc., 33, 5545 (1977); L. Carozzo, G. Corongiu, C. Petrongolo, E. Clementi, J. Chem. Phys., 68, 788 (1978); D. Ferro, M. Ragzzi, E. Clementi, J. Chem. Phys., 70, 1040 (1979); E. Clementi, C. Corongiu, J. Chem. Phys., 69, 4885 (1978); G. Corongiu, E. Clementi, E. Pretsch, W. Simon, J. Chem. Phys. 70, 1266 (1979).
15. H. Popkie, H. Kistenmacher, E. Clementi, J. Chem. Phys. 59, 1325 (1973); G. C. Lie, E. Clementi, J. Chem. Phys. 62, 2195 (1975); E. Clementi, G. C. Lie, M. Yoshimine, J. Chem. Phys., 64, 2314 (1976).

16. R. O. Watts, E. Clementi, J. Fromm, J. Chem. Phys., 61, 2250 (1974); E. Clementi, J. Fromm, and R. O. Watts, J. Chem. Phys. 62, 1388 (1975); R. Barsotti, E. Clementi, Theor. Chim. Acta 43, 101 (1976); E. Clementi, R. Barsotti, Chem. Phys. Letters, 59, 21 (1978).
17. S. Romano, E. Clementi, J. Inter. Quant. Chem., 15, 849 (1978).
18. E. Clementi, G. Ranghino, R. Scordamaglia, Chem. Phys. Letters 49, 218 (1977); G. Ranghino, E. Clementi, Gazz. Chim. It. 108(5-6), 157 (1978)
19. E. Clementi, G. Corongiu, B. Jönsson, S. Romano, FEBS 100, 313 (1979); E. Clementi, G. Corongiu, B. Jönsson, S. Romano, Gazz. Chim. It. (in press)
20. G. Corongiu, E. Clementi, F. Lelj, J. Chem. Phys. 70, 3726 (1978)
G. Corongiu, E. Clementi, Gazz. Chim. It. 108, 687 (1978)
21. E. Clementi, G. Corongiu, Biopolymer, 18, 2431 (1979).
E. Clementi, G. Corongiu, Chem. Phys. Letters, 60, 175, (1979).
E. Clementi, G. Corongiu, Gazz. Chim. It. 109, 201 (1979).
22. E. Clementi, G. Corongiu, J. Inter. Quant. Chem. 16(4), 897 (1979).
23. E. Clementi, G. Corongiu, Gazz. Chim. It., (in press).
24. G. Corongiu, E. Clementi, J. Chem. Phys. (submitted).

INDEX OF NAMES

INDEX OF SUBJECTS